Pirate Radio
and Video

Experimental
Transmitter Projects

Other Books of Interest by Newnes

Pirate Radio and Video

Experimental Transmitter Projects

Newton C. Braga

Newnes

Boston Oxford Auckland Johannesburg Melbourne New Delhi

ISBN 0-7506-7331-1

British Library Cataloguing-in-Publication Data
A catalogue record for this book is available from the British Library.

The publisher offers special discounts on bulk orders of this book.
For information, please contact:

Manager of Special Sales
Butterworth–Heinemann
225 Wildwood Avenue
Woburn, MA 01801-2041
Tel: 781-904-2500
Fax: 781-904-2620

For information on all Newnes publications available, contact our World Wide Web home page at:
http://www.newnespress.com

10 9 8 7 6 5 4 3 2 1

Printed in the United States of America

Contents

Preface

In the early days of electronics, all we had were radios. Electronics began with radios, and only after many years of emerging technologies, beginning with DeForest's triode, were we able to provide people with newer devices such as amplifiers and television sets and, more recently, pagers, digital television, and computers.

The traditional Radio Age, when people built their own receivers and transmitters, has practically ended, as the Internet and other new telecommunication technologies using satellites, fiber optics, and lasers have taken the place of home-made projects. Nevertheless, the idea that we can send a signal through walls without the use of any material media remains fascinating for us all. There is a level of mystery behind the operation of these devices.

But to learn how these wonders function, we must have a starting point. Building simple circuits that produce radio waves is the ideal point of departure for showing youngsters, students, beginners, teachers, and even researchers in other fields how the science of telecommunications works. That is the aim of this book.

Over the past 20+ years, the author has published many works directed toward students and beginners who want to learn basic electronics and has collected a large assortment of projects that involve radio and other wireless communications. Many of those projects are included herein. These are not the most advanced projects available. They were selected to be simple and accessible to all. However, many of them use modern components such as FETs and ICs.

Although the projects are basic, as they were created for experimental and didactic ends, we invite readers to improve them by creating new configurations or changing components to get better performance. Because they are educational and experimental in nature, these projects should prove very interesting to readers who want to explore the world of radio waves and wireless communications.

Many projects can stand alone as complete devices and be used for specific tasks in experiments and practical applications. Others can be employed as parts of more complex projects. If the reader is a student or a teacher, many of these projects will be useful in science fairs or as assignments in technology education at the high school and college levels.

It is necessary to warn the reader about the rules and laws controlling radio transmissions. The construction and operation of radio transmitters requires some care. Power and frequency levels must be careful chosen to avoid interfering with telecommunication services such as TV and radios used by neighbors. The content of the messages you intend to transmit must also be considered: commercials, political statements, and other messages that may be considered clandestine broadcasts in your locality must be avoided.

If the reader lives in the U.S.A., the FCC is the government bureau that controls radio emissions. Be sure that you do not infringe on any applicable rules or regulations while experimenting with your transmitter. And if the reader lives in another country, you must seek out information about the laws and rules that exist where you live so as to be in compliance when using the projects described in this book.

In Chapter 1, we discuss some techniques to help the reader understand how transmitters work and how to build them with a minimum of obstacles. We also provide some important information about their operation and how to avoid legal difficulties and interference problems with neighbors.

In Chapter 2, we describe an assortment of FM audio transmitters. The reader can use these projects to eavesdrop on conversations as part of a spy operation, to build an experimental FM radio station to broadcast home-made programs within a neighborhood or school campus, or to conduct demonstrations in science fairs or technology education.

In Chapter 3, we describe some special-purpose transmitters. Here, we include circuits designed to send information about temperature, the amount of light falling into a sensor, and other analog data to a remote receiver. We also include a wireless alarm and a transmitter that "jams" the signals of other transmitters by radiating electromagnetic noise.

In Chapter 4, we have an assortment of transmitters that operate in the MW to SW range. The MW range can be used for experimental radio stations at school or in the home, and the shortwave band can be used for experimental purposes.

Chapter 5 is dedicated to transmitters that use not only radio waves but other kinds of signals such as current fields conducted in the earth, magnetic fields, and laser beams. A TV transmitting station is also described in this chapter, along with a high-sensitivity transmitter for spies.

In Chapter 6, we describe complementary circuits that can be used with the transmitters. Those circuits include a field strength meter, power supplies, and a dip meter that can be useful for installing, testing, and adjusting transmitters.

It is important to remind the reader that, although many of these circuits are very simple and can be easily assembled and adjusted by inexperienced beginners, we nevertheless are working with high-frequency circuits. These circuits sometimes can be critical, and this should be considered when buying components, using nonrecommeded procedures, or using the circuit for purposes other than the ones indicated in the text.

Radio transmission is not only fun, it is a science. Readers who want to learn a great deal about this science are invited to begin with this book, moving to more advanced literature later. Of course, the first step begins here, with devices that are simple to build, using easy-to-find parts. Come with us to have fun and learn much about this fascinating science by building these transmitters.

Newton C. Braga

About the Author

Mr. Braga was born in São Paulo, Brazil, in 1946. He received his professional training at São Paulo University (USP). His activities in electronics began when he was 13 years old, at which time he began to write articles for Brazilian magazines. At age 18, he had his own column in the Brazilian edition of *Popular Electronics,* where he introduced the concept of "electronics for youngsters."

In 1976, he became technical director of the most important electronics magazine in South America, *Revista Saber Eletrônica* (published in Brazil, Argentina, and Mexico). He also has been the director of other magazines published by the same company, including *Eletrônica Total.* During this time, Mr. Braga has published more than 60 books about electronics, computers, and electricity, and thousands of articles and electronic projects in magazines all over the world (U.S.A., France, Spain, Portugal, Argentina, Mexico, et al.). More than 2,000,000 copies of his books have been sold throughout Latin America and Europe.

The author also teaches electronics and physics and is engaged in educational projects in his home country of Brazil. These projects include the introduction of electronics in secondary schools and professional training of workers who need enhanced knowledge in the field of electronics. The author now lives in Guarulhos (near São Paulo) and is married, with a 10-year-old son.

1

Useful Information about Transmitters

Transmitters are fantastic circuits. Just a few electronic parts mounted on a printed circuit board or terminal strip can send radio signals to a remote receiver without the need of wires or any other physical media between them. Also fantastic is the fact that the signal travels at the speed of light and is completely invisible to us. All of that makes radio transmission one of the most fascinating fields of electronics.

But an excess of enthusiasm when working in radio transmission devices can produce some undesirable results such as noise, interference, and other phenomena that can be seen by the authorities as legal transgressions. How the reader can go on the air with an experimental or small transmitter while avoiding problems with neighbors is the author's first preoccupation.

Knowing how to do this is important not only to readers who want to enter into the world of radio transmitters described in this book, but also to readers who want to adapt one or two of our projects to a particular application.

1.1   Producing Radio Waves

Electromagnetic waves (radio waves) are produced when a high-frequency current flows through an electric conductor. The waves are radiated to space and travel at the speed of light. A conductor that acts in this way, producing radio waves, is called an *antenna* or *aerial*.

Many people believe that the amount of radio signals produced by an antenna is associated with its dimensions: the longer the antenna, the more signal can be radiated into space. This is not true. The amount of energy transferred into space by an antenna is associated to the wavelength of the signal as related to the antenna dimensions (size and form) as shown in Fig. 1.1. The antenna's length is determined by its type, and it must maintain a numeric relationship with the wavelength.

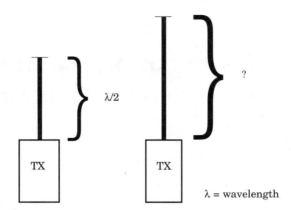

Figure 1.1   The wavelength and antenna length must be compatible for best performance.

Several types of antennas can be constructed for use with radio transmitters such as the ones shown in this book. All must have their dimensions calculated as function of the wavelength (and therefore the frequency) of the transmitted signal. An ideal antenna would transmit all of the radio frequency energy generated by the transmitter, as shown in Fig. 1.2. More energy transmitted into space translates into greater signal range.

But it is also very important to remind readers that more power doesn't always mean more range. The range of a transmitter depends on many factors other than power. This misconception may cause many readers to look for the "most powerful transmitter" to build for an application, in which case the results can be quite different from what is expected and cause great frustration.

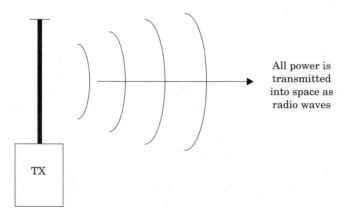

Figure 1.2   In an ideal system, all power produced by the transmitter is transmitted as electromagnetic waves.

Amateur radio operators can send signals distances of up to 1,000 miles using transmitters with power outputs less than a few milliwatts. Many of those very low-power transmitters produce less power than the smallest transmitter described in this book, whose signals can't travel beyond a few feet under normal conditions. To obtain those long-range transmissions, radio amateurs use special techniques involving directional antennas, careful choice of frequencies, favorable propagation conditions, and so on.

You might also consider that TV stations may use transmitters with powers in the range of 100,000 W or more, and yet their signals can't be picked up by receivers located at distances beyond 150 or 200 miles. Why is this?

The operating frequency, the position of the transmitter antenna and the receiver, and topographic conditions (the presence of hills, valleys, forests, and large buildings) are factors that can affect the distance covered by the radio signals from a transmitter. And, depending on the frequency, other factors must be considered when we study transmission range.

Signals at frequencies above 30 or 40 MHz, under normal conditions, don't reflect or bend when they reach the ionosphere. But signals at lower frequencies (referred to as *shortwave* or *medium wave*) reflect off the ionosphere and also off the ground. Via successive reflections or bending, they can cover very large distances.

Figure 1.3 shows how a shortwave signal can travel long distances by successive reflections between the ionosphere and the Earth. This is why you can tune in weak shortwave radio stations on the other side of the world, but you can't re-

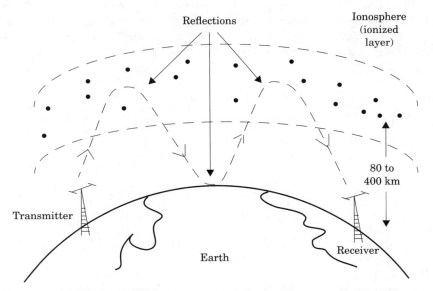

Figure 1.3   Shortwave signals can cover long distances by successive reflections between the ionosphere and Earth.

ceive signals from a powerful TV station in a city that is only a few hundred miles away. They use different kinds of signals.

Another factor to be considered is the dimension of an obstacle when compared with the wavelength of a radio signal. Large-wavelength signals as the ones produced by MW and LW (medium wave and long wave) stations can be picked up at great distances because they can pass through large obstacles such as hills and buildings. But shortwave signals such as the ones produced by FM and TV stations are blocked by those obstacles. The reader can observe that is easy to pick up an MW station behind a hill, but not FM stations or TV stations, as shown in Fig. 1.4.

If a transmitter can send its signals over very large distances (e.g., in the shortwave range), we have to consider an additional factor, very well known by radio amateurs (hams): *propagation*. Radio amateurs know that they can't send their signals to any desired location at any moment, no matter how powerful the transmitter. You cannot switch on even a very powerful transmitter and expect to talk with someone in a remote region of the world without any problems.

The conditions under which the signals can reflect in the ionosphere change according the day time and the seasons of the year, and reflection is also affected by the radiation produced by the sun (magnetic storms).

Depending on the time, and according to the ionosphere's position and density, the signal produced by a shortwave transmitter might be picked at distances up to 1,500 to 2,000 miles but not by a receiver placed only several miles from the station (Fig. 1.5).

Radio operators consult tables of propagation conditions before they attempt to make contact with a specific geographic location. They must be sure that contact is possible before turning on their equipment and sending a message. It is pointless if the intended recipient cannot receive the signal.

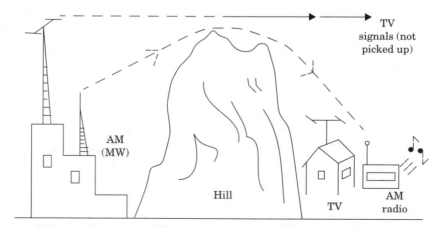

Figure 1.4   AM (MW) radio station can be tuned behind large obstacles.

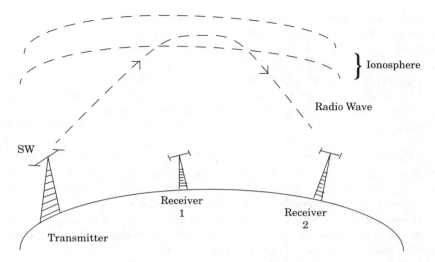

Figure 1.5   The signal can't be tuned in by receiver 1, even though it is closer to the transmitter than receiver 2.

Taking into consideration all of the above, the reader can understand that it makes no sense to say, "I want a transmitter that can cover *x* miles." This is particularly true in the shortwave range. Transmitters with 0.001 W or 1,000 W of power can send their signals to the same place—depending on all related factors. Of course, if conditions are favorable, two transmitters placed at the same point but having different output powers will show different signal strengths at the reception point.

This kind of reasoning can be applied to small transmitters operating in the range of 30 to 300 MHz. Low-power signals, when propagating through a medium without obstacles, can reach enormous distances. Using only 25 W, a transmitter aboard the space probe *Mariner* sent signals from Mars to Earth, covering a distance of about 40,000,000 miles! And if you install a parabolic reflector to tune TV signals from a satellite in orbit 20,000 miles away, you will be receiving signals that were sent by a transmitter that used only a few watts of power.

Considering this, the reader can see that our small transmitters, correctly designed, using appropriate antennas, and placed in favorable points, can reach distances much greater than the ones indicated in the projects as *typical*. It is up to the reader to determine the best way to send the signals the greatest possible distance.

1.2   Frequency

In its simplest version, a transmitter is no more than a high-frequency oscillator that produces a radio signal. The signal is nothing more than a high-frequency electric current flowing in the circuit, and it can be applied to an antenna to produce radio waves as previously described. In a small transmitter such as this, the transistor characteristics determine the highest frequency that can be produced and the highest available power.

Figure 1.6 shows two typical circuits. From these we can see that, if we increase the supply voltage, we can get more power from the oscillator. But, at the same time, the transistor produces more heat. Because there is a limit to how much heat a transistor can handle, there is a corresponding limit to how much power we can generate with this circuit.

In addition, when we increase the input power, we also must consider that the gain of the transistor decreases as more power is dissipated as heat, so the efficiency of the process is reduced. Operating near its overload limit, the transistor converts much more power in heat than into radio waves.

If we need to get more power from a transmitter, it is convenient to separate the wave generation functions to increase the power of the signals. Powerful transmitters use a low-power transistor to oscillate and generate the high-frequency signal, and another transistor in another stage to increase the signal power and apply it to the antenna, as shown in Fig. 1.7.

But this is a more complex circuit that needs special care with regard to the layout of the components when they are assembled, and the specific component characteristics must be chosen according to the application. Special transistors designed for high-frequency amplification must be used in critical projects.

Each pair of coils forms a transformer, and their function is to transfer the signal from one stage to another without power loss. This means that these coils must be carefully tuned to the operating frequency.

If you need to build a professional-level transmitter, you also have to use special transistors designed for high-frequency or radio communication applications. The transmitter needs to have the ability to amplify high frequencies with small losses and high gain, among other important characteristics.

For our reader, it is important to know that, in some cases, audio transistors and common high-frequency transistors used in FM and TV receivers also can be used successfully in the experimental transmitters described in this book. This is of interest because it indicates that you can obtain these transistors easily from any electronics surplus store or component dealer, and they are cheap.

The reader must understand also that, as the signal frequency increases, the transistor's ability to work with these signals decreases. There is a point at which the transistor gain is reduced to 1, and no amplification occurs; the input signal has the same amplitude as the output signal.

In this condition, the transistor is not useful in any practical applications. The highest theoretical frequency at which a transistor can act as an effective amplifier (i.e., the gain > 1) is called the *transition frequency* and is indicated as f_T in technical publications, data sheets, and application notes. Observe that we can find cheap transistors with high power capacities (e.g., the 2N3055), but they have a very low f_T (e.g., 1 MHz), which limits their application to LW and MW transmitters.

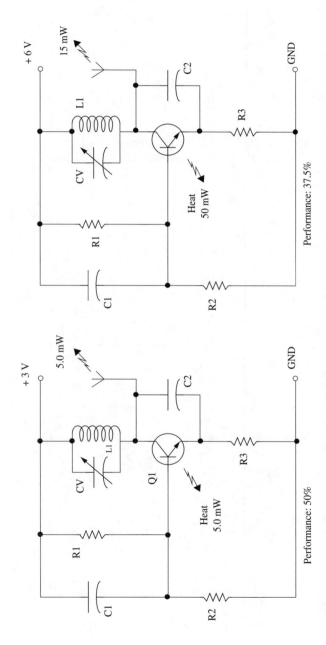

Figure 1.6 Increasing the power supply voltage, the performance decreases.

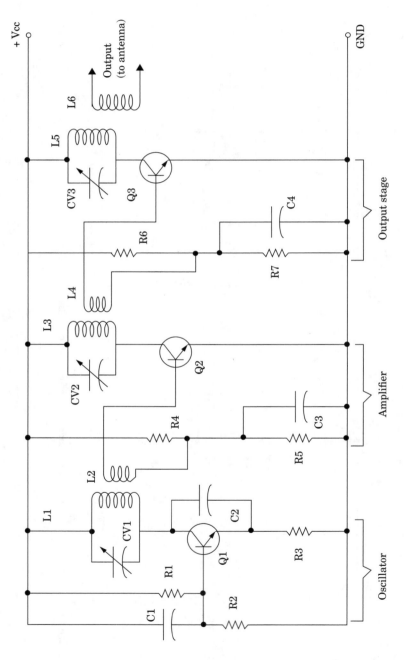

Figure 1.7 A three-stage transistor transmitter.

Many radio amateurs use high-power stages or amplifiers to increase the power of their transmitters. The *linear amplifier,* as the used device is called, is plugged into the output of a transmitter to increase its power.

1.3   The Law

Nearly every transmission of radio waves is controlled by federal laws and regulations. In the U.S.A., the Federal Communications Commission (FCC) controls all radio transmissions and conducts a continual inspection of the radio spectrum. Sensitive receivers are used to detect any "suspect" signal that might interfere with legal radio communications services, and these receivers can determine the transmitter's location.

A "pirate transmitter" can be detected and located in a few moments, and the persons responsible for its operation are subject to arrest and prosecution. It is therefore important to understand the basic boundaries of telecommunications law, as some of our transmitters, if used improperly, can infringe on these laws. It is not our intention to send the reader to jail . . . .

Of course, the small transmitters that use only one or two transistors cannot harm anyone, as their signals are restricted to the reader's home or other relatively small areas. But if they reach your neighbor's TV or FM receiver and cause interference, you can have problems. We can't predict if your neighbor is a cranky person or simply doesn't like you . . . .

1.3.1   FCC Rules

If you live in the U.S.A., it is important to understand the FCC regulations that must be observed when putting any kind of transmitter (even small ones) on the air. If you reside in another country, it is important to obtain this information from local regulatory agencies.

1.4   Transmission Techniques

Assembling a small transmitter is a simple task, but there are some important considerations when the circuits need some kind of adjustment or when special parts or layouts are required. This section is dedicated to the description of some basic transmission techniques and to providing useful information for the reader who is not experienced with this kind of project.

1.4.1   What a Transmitter Is

The existence of radio waves was predicted by Maxwell, but they were produced in the laboratory only many years later (1887) by Heinrich Hertz, and subsequently by many other researchers, including Popov in Russia, Marconi in Italy, and Landel de Moura in Brazil. The discoveries made by these scientists and others formed the starting point from which we have derived modern technologies such as cellular telephones, satellite communications, pagers, etc.

To understand how a transmitter works, it is important to start with the nature of radio waves. When a electric current flows through a wire, electric and magnetic fields are created around it, producing a *perturbation* that is radiated into space at the speed of light (186,000 miles or 300,000 kilometers per second). Thus, the wire acts as a radiating *antenna*. Unlike sound waves, this perturbation or *electromagnetic wave* doesn't need any material (or *medium*) to support its propagation. That means that radio waves can travel through a vacuum.

It is quite important in telecommunications that radio or electromagnetic waves can be used to carry information from one place to another without the need of any physical medium, such as copper wire. To receive the waves in a remote location, we only have to use another electric conductor (the receiving antenna) to intercept those waves and produce an electric current. We then use a special device called a *receiver* to convert this current into useful information, as shown in Fig. 1.8.

By appropriately changing the electric current produced by a transmitter, it is possible to send useful information to the receiver in the form of coded messages (Morse or telegraphic code), voice and music as in broadcasting, images as in TV, data from a PC or a sensor, etc.

So, a transmitter is made up of two basic sections:

1. An electronic circuit that produces high-frequency currents, which contain information to be sent to a distant location
2. An antenna system to which the currents are applied to generate radio waves

A simple transmitter is formed by a high-frequency oscillator (using one transistor or one tube) and an antenna as shown in Fig. 1.9. Starting from this simple configuration, a transmitter can become increasingly complex, depending on fac-

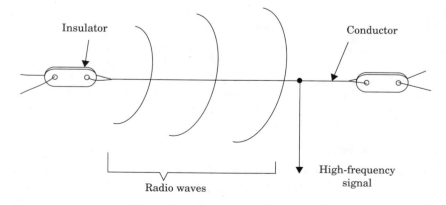

Figure 1.8   Any conductor intercepted by radio waves acts as an antenna.

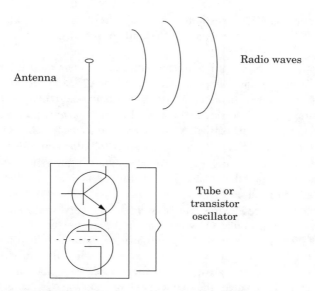

Figure 1.9   The simplest transmitter is formed by an oscillator and an antenna.

tors such as the desired output power, the kind of information to be sent, and the frequency range.

Figure 1.10 shows a transmitter that includes several special stages. This is similar to the ones found in broadcasting and telecommunication stations or ama-

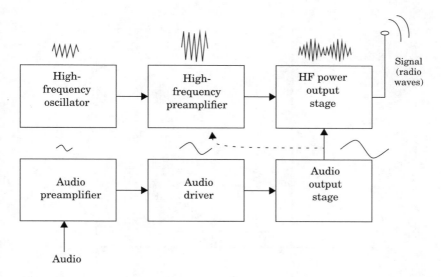

Figure 1.10   A transmitter using several audio and HF stages.

teur radio stations. This transmitter has a high-frequency stage where the high-frequency (HF) or radio frequency (RF) oscillator produces a signal at the chosen frequency.

For frequency stability, professional-quality transmitters must use piezoelectric crystals or quartz crystals (simply called *crystals* or *XTALs* by technicians). Quartz crystals are components that tend to oscillate at only one frequency when excited by an electric signal. The size and shape of a piece of quartz crystal inside the component determines its particular oscillation frequency, which does not vary.

Figure 1.11 shows a simple transistorized oscillator controlled by a quartz crystal. This circuit can be used to produce signals in the range between 2 and 20 MHz, depending on the crystal. The stability of a circuit like this can be measured in parts per million (ppm).

The signals produced by this type of oscillator are very weak and can't be used in a long-range transmitter. Only experimental transmitters with ranges up to a few hundred feet can be based on a one-transistor configuration or a simple high-frequency oscillator. To get more power, we must apply the generated signal to amplifier stages. In some cases, the power amplifier stage not only increases the signal power, it also changes the frequency—for instance, by a factor of two. This kind of amplifier stage is also called a *frequency doubler.* Some transmitters have only one amplification stage, but some of the larger ones use two or more stages to increase the power to very high values.

Up to this point, the signals found in our circuits are merely high-frequency currents that do not contain any kind of information. The information to be car-

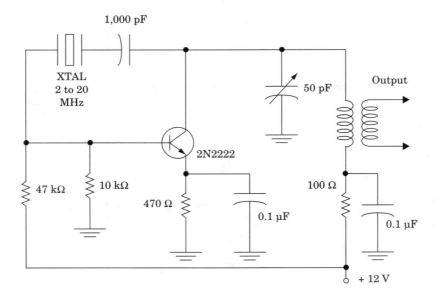

Figure 1.11   A crystal (XTAL) controlled oscillator.

ried by the radio signals must be processed by other stages. Those stages alter the high-frequency signals, producing what we call *modulation*.

The high-frequency signal that carries the information is called the *carrier*. So, we need to *modulate a carrier* if information is to be sent by a transmitter.

In the case of sound (music and voice), the modulation circuits are simple audio amplifiers. We have a preamplifier to pick up the signal from a microphone, tape recorder, or CD player; an intermediate amplification stage (drive); and a power output stage, exactly as found in amplifiers for domestic use.

The power needed to produce the correct modulation of a high-frequency signal depends on the type of transmission. In the next section, we will discuss this subject, but it generally is in the same range as the transmitted signal.

Some small amplifiers do not need much power to perform the modulation process, so a simple stage can be used for this task. In this book, for instance, we will show some small transmitters in which the modulation stage doesn't exist or uses only one transistor.

A high-power modulation stage is needed to produce a 100% modulated signal, which is required for improved transmitter performance. If we have a very high-power transmitter with an output in the range of several kilowatts, such as used by broadcast stations, it is common to modulate the signal by one or two stages before the final stage, as shown in Fig. 1.12. Using this kind of circuit configuration, it is possible to achieve modulation of a 100 W transmitter from a small 5 W audio amplifier.

Several modulation techniques exist, and they will be described next. The important fact for the reader to keep in mind is that the number of RF stages and modulation stages depends on several factors and varies from transmitter to transmitter.

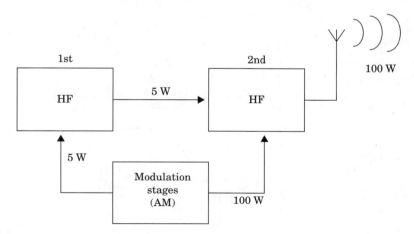

Figure 1.12 Less power is required for modulation if applied to the first high-frequency stage.

1.4.2 Modulation

The simplest way to use a transmitter to send information using radio signals is by continuous wave (CW). The transmitter is a simple high-frequency oscillator plugged into an antenna as shown in Fig. 1.13.

In this kind of transmission technique, the signals produced by a circuit are switched on and off by a key. A short signal burst represents a dot (·), and a long signal burst is a dash (–). Combining dots and dashes in Morse code, we can represent all alphabetic and graphic symbols and numbers. (See the appendix for a listing of the complete Morse code.) For instance, a short burst followed by a long burst [coded as a dot-dash (· –)] means the letter A.

Experienced operators can send complete, long messages very quickly. And, of course, it is necessary to have a good ear to decode the received signals.

Today, this kind of transmission is used only by a few operators. (Radio amateurs and radio operators must know Morse code, as it can be useful in emergency situations.) You can tune in this kind of transmission with your shortwave receiver (in the 7 to 14 MHz range, for instance) as long and short or burst of signals without any modulation.

It is important to remind the reader that a CW transmitter is so simple that, in an emergency situation, even an antique radio receiver for the MW or SW band, or a television set, can be converted to send telegraphic signals.

Figure 1.14 shows an audio output stage as used in many antique radios and TVs using a pentode tube. By replacing some components wired to this tube, as shown in the same figure, we can change it into a CW transmitter for emergency situations. A transmitter like this can send signals distances of hundreds or thousands of miles under favorable propagation conditions.

To send signals in the frequency range between 3 and 7 MHz (40 and 80 meter radio amateur band), the coil is formed by 15 + 15 turns of enameled AWG 28

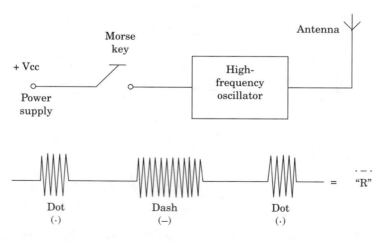

Figure 1.13 A continuous wave (CW) or telegraphic transmitter.

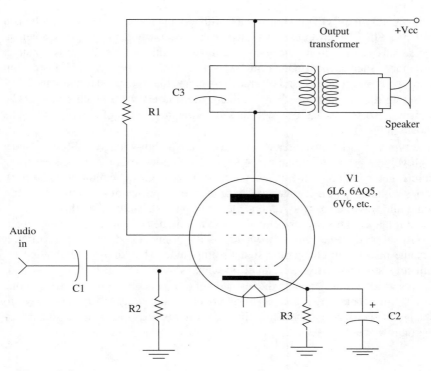

(a) audio output stage

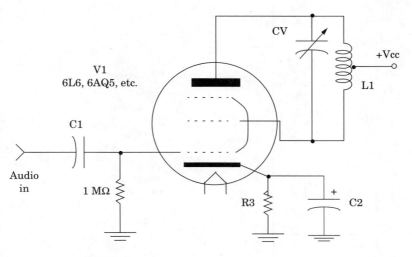

(b) converted to a transmitter

Figure 1.14   Converting the audio output stage of an old tube radio receiver into a transmitter.

wire wound on a PVC or wooden tube with a diameter of 1 inch. If the tube is a 6V6 or a 6L6, the power can be as high as 5 W and, with an appropriate antenna, the signals can reach distances up to a thousand miles. Less power is obtained from tubes such as the 6AQ5 or 50C5. Using the transmitter is very simple: using a key, you merely turn on and off the high voltage that is applied to the stage.

Before the vacuum tube was invented, transmitters were much simpler. The transmitters used at that time were based on the production of electric sparks, as shown in Fig. 1.15.

A typical spark transmitter was formed using a high-voltage coil to produce a high voltage between two points. A spark was produced in the gap between the points, generating a high-frequency signal. A resonant LC circuit was placed in series with the spark generator to determine the frequency of the signal. Marconi sent radio waves across the Atlantic Ocean using a transmitter like this.

With the invention of tubes, modulation techniques advanced. Injecting an audio signal produced by a low-frequency oscillator into a high-frequency carrier, it became possible to change the signal's amplitude. This kind of modulation results in a tone-modulated telegraphic system, as shown in Fig. 1.16.

We can also consider using the injection of a signal coming from a microphone to alter the characteristics of a high-frequency carrier. There are two ways to change the carrier characteristics and therefore two ways to transmit sounds using radio waves.

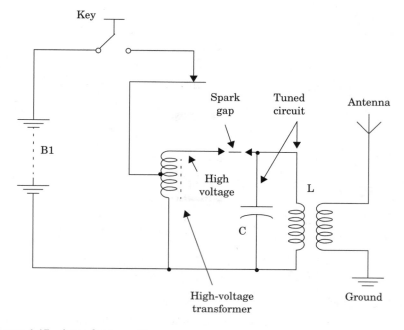

Figure 1.15   A spark transmitter.

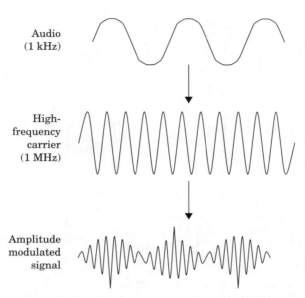

Audio
(1 kHz)

High-
frequency
carrier
(1 MHz)

Amplitude
modulated
signal

Figure 1.16   The amplitude of a carrier changes with the low-frequency signal in AM.

Amplitude Modulation.   Using amplitude modulation (AM), we can change the amplitude or the amount of power of the radio signal using the low-frequency or audio signal. For instance, if we have a 1,000 Hz signal to be transmitted, this signal changes the amplitude of the radio signals 1,000 times per second.

To be more efficient, the high-frequency signal in a radio transmission using amplitude modulation must be 100% modulated. This means that the amplitude of the carrier can vary in a range of 0 to 100% of its value according to the audio signal. If the changes are less than this value, we have a condition called *under-modulation,* and the efficiency of the transmitter is low. If we create changes above 100%, as shown in Fig. 1.17, an *overmodulation* occurs, causing problems relating to interference, distortion, and undesirable signals produced by the equipment.

Amplitude modulation is used in broadcasting stations operating in the MW and SW ranges, by radio amateurs, and in many other telecommunication services. It has some disadvantages, one of which is sensitivity to noise and interference. Noise and interference signals can appear at the moment when the radio signal is in the lowest amplitude point in each cycle. That means that they prevail and appear as noise reproduced by the receiver's loudspeaker.

Frequency Modulation.   Another kind of modulation is called *frequency modulation (FM).* To produce an FM signal, we must change the frequency of a carrier (not the amplitude) according to the amplitude and frequency of an audio signal

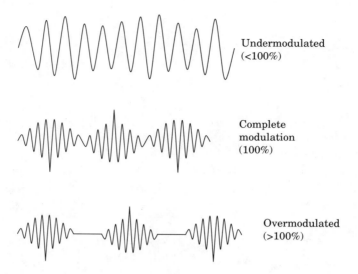

Figure 1.17 Amount of modulation to an AM signal.

as suggested by Fig. 1.18. For instance, if we use a 1 kHz audio signal to modulate a 100 MHz signal, the 100 MHz will change its frequency between two values (for instance 99.5 and 100.5 MHz) 1,000 times per second, according to the audio signal.

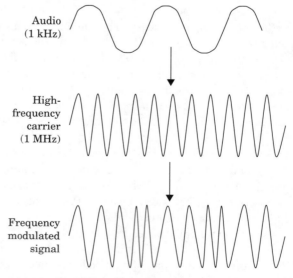

Figure 1.18 The carrier's frequency changes according to the audio signal (the amplitude remains constant).

The principal advantage of this kind of modulation is that we have no low-amplitude points to be affected by noise and interference. This means that FM is less sensitive to noise than is AM.

1.4.3   Oscillators

Several circuits can be used to produce high-frequency signals. The circuits applied to this task are called *oscillators*, and their configurations depend on the desired frequency, power, and many other factors.

Let's examine some important oscillators (many of them used in projects described later in this book). The first oscillator to consider is the one shown in Fig. 1.19. This one uses a quartz crystal. In this circuit, the crystal determines the operation frequency, and the coil acts as a load to the produced signal, allowing it to be transferred to the next stage.

This circuit can be used to generate signals in the range between some hundreds of kilohertz to more than 50 MHz. The variable capacitor is used to find the best oscillation point and maximum performance when transferring the signal to the next stage.

Figure 1.20 shows another oscillator that uses a crystal to fix the frequency. This circuit is called a *Pierce oscillator,* and it uses a field effect transistor (FET). Frequencies up to some tenths of a megahertz can be generated by this circuit. As the circuit uses a crystal, the only way to change its frequency is by replacing this component by another that will produce the new desired frequency.

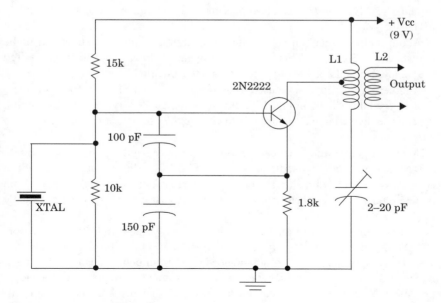

Figure 1.19   A crystal-controlled oscillator.

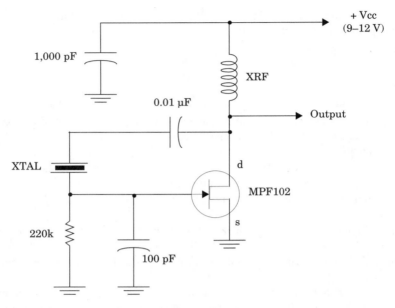

Figure 1.20   Crystal-controlled Pierce oscillator using an FET.

An important characteristic of this circuit is that it can oscillate in an *overtone*. This means that if we use a 7 MHz crystal, the circuit can be adjusted not only to produce a 7 MHz signal but also signals at multiple frequencies such as 14 and 21 MHz.

Another circuit that can be adjusted to operate at multiple frequencies is the one shown in Fig. 1.21. This circuit must include a filter to separate the desired frequency from others that are produced at the same time.

Another circuit that can be used to produce 7, 14, and 21 MHz signals using a 7 MHz crystal is shown in Fig. 1.22. It is important to observe that, in this circuit, we can use the variable capacitor CV to change the oscillation frequency within a narrow range.

A very popular circuit that can be used to produce signals at frequencies up to 100 MHz is the Hartley oscillator shown in Fig. 1.23. This circuit can be built in two configurations: with a transistor or with a triode tube. Notice that the coil can be wired both to the collector and emitter of a transistor, or to the plate or cathode of a tube.

The circuits shown in Fig. 1.23 can be used as variable frequency oscillators (VFOs). As opposed to the crystal oscillators, a VFO can produce signals in a wide range of frequencies without the need of any component changes. The resonance of the coil and the variable capacitor determine the frequency of the signal produced. Notice that, in this circuit, the feedback that keeps the circuit in oscillation is produced by a tap in the coil.

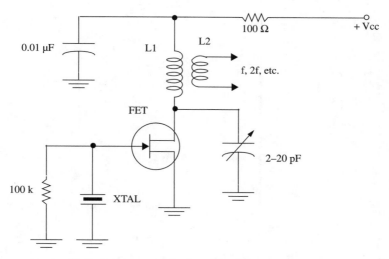

Figure 1.21   Oscillator/frequency multiplier using an FET.

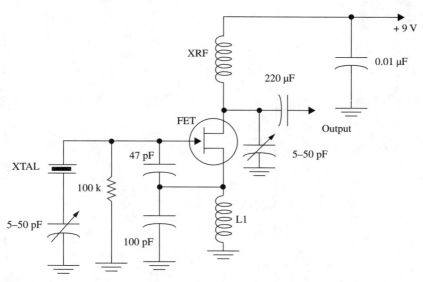

Figure 1.22   Oscillator for the frequency range between 14 and 21 MHz using a 7 MHz crystal.

In the Colpitts oscillator, the feedback is made by capacitive derivation (not inductive, as in the Hartley), but the operating principle is the same. Figure 1.24 shows a Colpitts oscillator that can be used to produce signals up to 300 MHz. The circuit shown uses an FET, but there are versions that use bipolar transistors and also tubes. The reader can observe that the transistor is wired in a common base configuration.

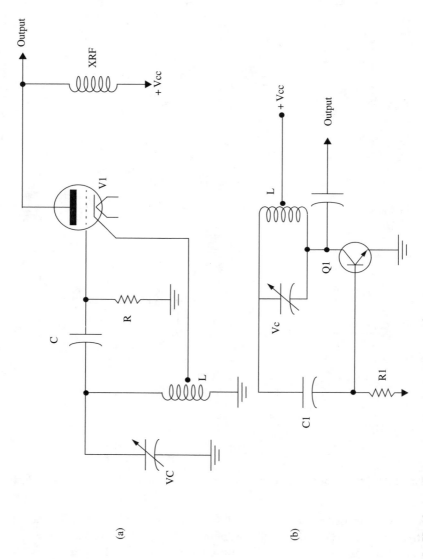

Figure 1.23   Hartley oscillator using (a) tube and (b) transistor.

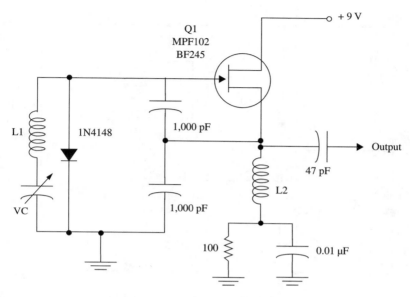

Figure 1.24   Colpitts oscillator using an FET.

1.4.4   Oscillator Output Power

High-frequency oscillators can be used to produce signals with power outputs ranging from several milliwatts to several watts. The signals in the range of several watts are produced by circuits using power transistors or tubes. This means that some of these oscillators can be used alone in small, short-range transmitters.

Depending on the intended use, it may be enough to modulate the high-frequency signal to arrive at a complete low-power transmitter. A small transmitter using only one transistor as high-frequency oscillator and modulated by a microphone can be seen in Fig. 1.25.

The reader will find complete data about this transmitter in the practical application chapters of this book. But it is useful to provide some advance information at this point.

In this circuit, the coil is formed by four turns of AWG 22 wire on a form (air core) with diameter of 1 cm. The circuit will send signals in the FM range between 88 and 108 MHz. Using a 6 V supply, the signals can be picked up at distances up to 300 ft. The antenna is a piece of wire 10 to 40 inches long.

Using powerful transistors such as the 2N2218 or BD135, and powering the circuit from 9 to 12 V supplies, it is possible to increase its performance. The signals then can be picked up at distances up to half a mile.

An equivalent circuit can be made using tubes such as the 6C4 or EL84. The distance covered by the signals in this case can extend up to many miles, depending on the antenna and the factors discussed previously.

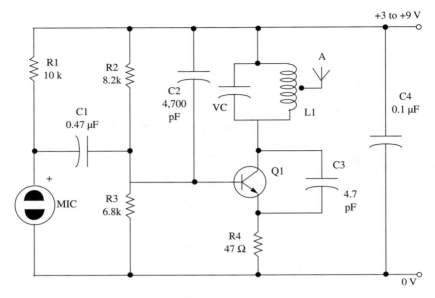

Figure 1.25 Small one-transistor FM transmitter.

The circuit shown in Fig. 1.26 produces strong signals from the SW range up to 200 MHz, depending on the transistors used. You can tune in the signal produced by this circuit at distances up to many hundreds miles if you use an appropriate antenna and under favorable propagation conditions. To operate this circuit as an experimental medium wave (MW) radio station (e.g., in school or a club)

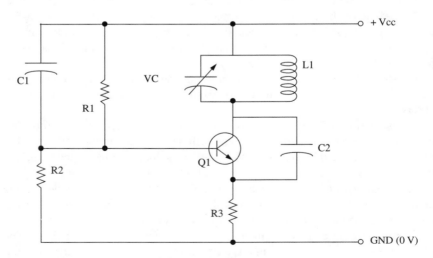

Figure 1.26 This circuit can produce signals up to 250 MHz.

the coil is formed by 50 + 50 turns of AWG 28 wire on a ferrite core (1 cm dia. and 20 cm long), and the transistor should be the TIP31 or the 2N3055. A 12 to 15 V supply is used to drive this circuit.

1.4.5   High-Frequency Amplifiers

A power output stage can increase the power of a transmitter. A power output stage can be internal (if part of the original circuit) or external (if plugged into the output of a transmitter).

External power amplifiers used to increase the performance of a transmitter are called *linear amplifiers* and are very popular among radio amateurs. They can process signals in a range between 0.5 and 100 W, resulting in output signals in the range of 50 to 2,000 W.

Figure 1.27 shows a diagram of a simple linear amplifier that can be used with small FM transmitters. The 6C4 tube is not an RF tube, and it can produce outputs up to only 4 W, but it is cheap and even can be found in old radio and television sets.

An important factor to be considered when projecting a power output stage is the class of operation. We can see in Fig. 1.28 that, depending on the bias of a stage, the signal can be cut at different amplitude points. The point is determined by the instant at which the device used in the stage becomes active or when the current flow begins.

In Class A, the device (transistor or tube) is biased so that both signal half-cycles are amplified. This means that the current flows through the device all the

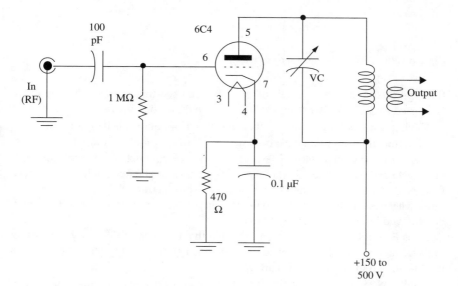

Figure 1.27   A simple linear amplification stage to the VHF range.

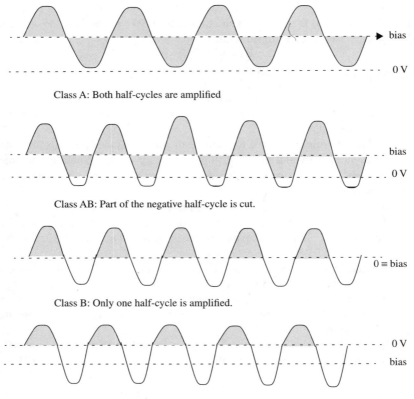

Class A: Both half-cycles are amplified

Class AB: Part of the negative half-cycle is cut.

Class B: Only one half-cycle is amplified.

Class C: A small part of the positive half-cycle is conducted.

Figure 1.28   Amplification stages by class.

time, producing a large amount of heat. We can see that most of the energy is converted into heat rather than used to increase the signal power.

The highest theoretical gain for this circuit is 50%. For each 10 W of power applied to the circuit, we can have only up to 5 W of power at the output signal.

Biasing the output stage in Class AB, the device conducts (and amplifies) one half-cycle and part of the other as shown in the figure. The net result is better performance, but the cut of a half-cycle causes the production of undesirable oscillations. These oscillations can cause interference if they are radiated, and they must be removed by the use of special filters.

Class B is another configuration in which the device amplifies only one of the high-frequency signal's half-cycles. The gain is about 60%, but the circuit is also nonlinear, and many harmonics are generated.

Finally, we have Class C. Biasing the device in this class, it conducts only a part of one half-cycle. The gain is very high, about 80%, as the input power is almost all used to produce the output signal. It is also necessary to use a filter to eliminate undesirable signals produced by this configuration.

Some examples of circuits using the described configurations are shown in Fig. 1.29. An important configuration based in the simple Class B biasing is the one called the *push-pull* stage. This configuration uses a two-transistor Class B bias as shown in Fig. 1.30.

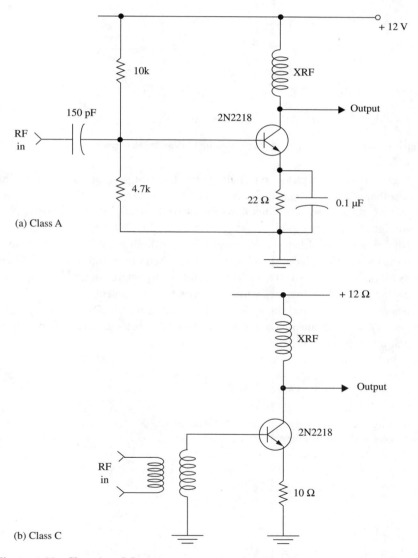

Figure 1.29   Class A and C output stages using a medium-power low-cost transistor.

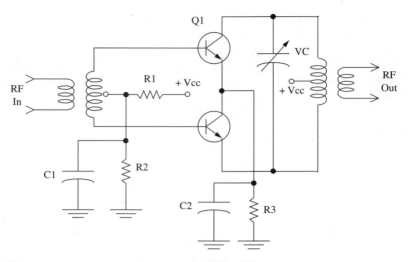

Figure 1.30   A push-pull output stage using bipolar transistors.

Each transistor amplifies one half-cycle of the high-frequency signal. This means that when one transistor is conducting, the other is cut off. The result is a high performance for the circuit, which converts nearly 100% of the applied power into the output signal.

Note that the amplifier stages can present two tuning configurations. One of them is the *aperiodic* configuration, so called when the circuit has no tuned circuits placed in the output as load. This circuit is very interesting, as it can operate in a wide range of frequencies without the need of adjustment.

Another configuration, shown in Fig. 1.31, uses a tuned LC circuit at the stage output. This circuit must be adjusted carefully to be resonant at the transmitter's operation frequency.

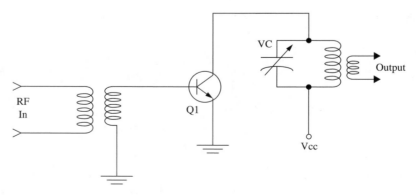

Figure 1.31   Tuned load output stage.

1.4.6   Output Circuits

To transfer all the power produced by the transmitter to the antenna, it is necessary for both impedances to be equal. This means that, to get better performance, it is necessary to ensure an impedance match between the antenna and the transmitter. Matching circuits are important not only to transfer all the power to the antenna and into the air but also to reduce the level of undesirable signals. A simple impedance matching network used with small FM and VHF transmitters is shown in Fig. 1.32.

The transmitter output circuit has a high impedance as compared to the antenna impedance. Therefore, we have to wire the antenna to a tap in the coil to reduce the impedance.

This kind of coupling technique is important when using telescoping antennas, as their impedance depends on their length. If the telescoping antenna is wired directly to the transistor collector, the amount of power transferred into space as radio signals is less than the power generated by the circuit. The difference between the produced output power and the actual power transferred into the space indicates a loss to be avoided. There is also the problem of instability; antenna oscillation or the act of placing the antenna near your hand or another object is enough to affect the frequency, detuning the transmitter. Depending on the antenna length, the reader will need to experiment to find the number of turns in the coil that produce the best performance.

Another way to couple a transmitter output to an antenna is shown in Fig. 1.33. A second coil or secondary winding, if we consider the network as a high-frequency transformer, is used. The necessary number of turns to match the circuit with the antenna is found from experimentation, or it can be calculated from the formula shown in Fig. 1.33. In the formula, X_L is the circuit impedance in ohms,

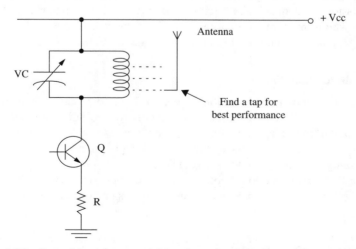

Figure 1.32   Correct impedance matching depends on the tap used to connect the an-

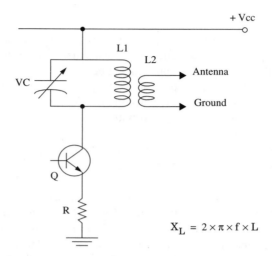

Figure 1.33   Coupling an output stage to a low-impedance antenna.

f is the operating frequency in hertz, L is the coil inductance in henries, and π is the constant 3.14. Calculating the circuit impedance and knowing the antenna impedance, the reader can determine the proper turn ratio between the coils.

Figure 1.34 shows an important configuration used to match impedances while reducing harmonic emissions. A second harmonic emission reduction up to 50 dB can be obtained with a circuit such as this. The circuit must be correctly turned to achieve the best results.

Another configuration using high-frequency transformers such as the *balanced-to-unbalanced (Balun)* device can also be used to reduce undesirable signal emissions and match the circuit impedance with the antenna impedance. Readers who want more information about this subject, and about radio transmission in general, can find useful information in the *ARRL Handbook.*[*]

1.4.7   Modulation Circuits

When using amplitude modulation (AM), we have to produce changes in the level of the high-frequency signal corresponding to the audio signal or low-frequency signal (music or voice). Several methods are available to do that.

Using tube circuits, we have three possible configurations to achieve AM. These configurations are shown in Fig. 1.35.

In the first circuit, the low-frequency signal is applied to the output tube control grid (a). For this configuration, the tube can be a pentode or a tetrode. The au-

---

[*] *ARRL Handbook for Radio Amateurs,* 77th edition, 2000, American Radio Relay League, 225 Main St., Newington, CT 06111.

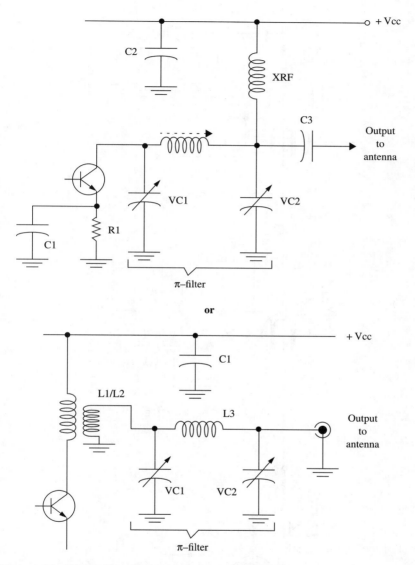

Figure 1.34   Output filters used to reduce harmonics irradiation.

dio signal controls the current flow between anode and cathode, changing the amount of high-frequency power in the tube output. This configuration's principal advantage is that the amount of power necessary for a complete modulation is less than the stage output power.

In another configuration, shown as part (b) of the same figure, the low-frequency signal is applied to the cathode circuit. A transformer must be used to

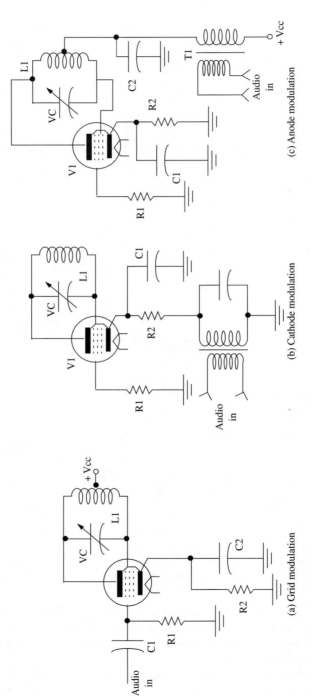

Figure 1.35   Amplitude modulation circuits using tubes.

couple the audio output stage to the final transmitter stage. When using this circuit, the amount of power needed to get a complete modulation is the same as the transmitter's output power. Finally, in part (c), we show the configuration where a transformer is used to control the high voltage supplied to the circuit.

An important circuit used for amplitude modulation, shown in Fig. 1.36, can be added to the ones shown previously. Another tube is used to control the amount of current through the transmitter output stage from the audio signals. The power managed by the two tubes should be equal.

Observe that it is necessary to wire a capacitor between the cathode and the ground of these circuits to decouple the high-frequency signals and avoid their presence in the low-frequency circuits.

Of course, all these configurations have transistor-based equivalents as shown in Fig. 1.37. In this case, we must consider that transistors have only one control element instead of two or three, as found in the tetrode or pentode tubes.

Frequency modulation (FM) can also be achieved using several configurations. For small, experimental transmitters, where the applications are not critical, FM can be created by changing the oscillator transistor biasing as shown in Fig. 1.38.

The process can be easily explained: when we change the base current, we also change the capacitance between base and emitter, which affects the oscillation frequency. So, an audio signal applied to the base of a transistor wired as oscillator, as shown in the figure, produces an FM signal.

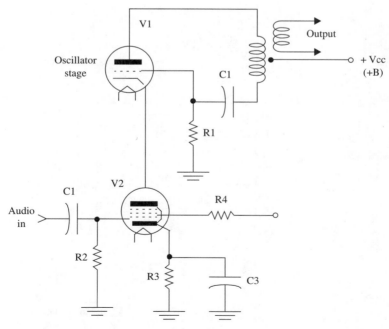

Figure 1.36   Tubes wired in series to perform an amplitude modulation stage.

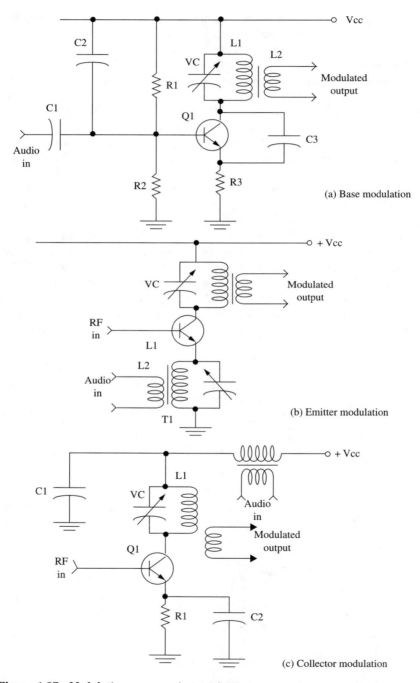

(a) Base modulation

(b) Emitter modulation

(c) Collector modulation

Figure 1.37   Modulation processes in transistor stages.

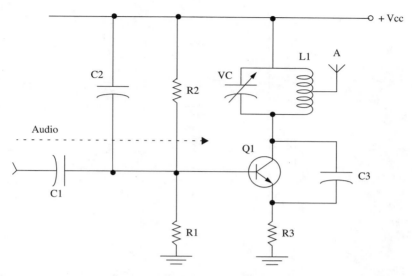

Figure 1.38   A simple way to produce FM signals using an oscillator.

The principal advantage to be considered in this configuration is simplicity: a high-impedance microphone can be wired directly to the base of a transistor. The same effect can be observed if a low-impedance microphone is wired to the transistor's emitter.

For critical applications, we have special configurations using such components as variable-capacitance diodes or *varicaps*. These diodes, when reverse biased, show a capacitance that depends on the applied voltage. (See the "Powerful Varicap Modulated Transmitter" project described elsewhere in this book.) A frequency modulation circuit using this component is shown in Fig. 1.39.

The audio signal is applied to the diode, passing by a resistor and a coil. (The coil prevents the high-frequency signal from reaching the audio circuit.) The changes in the voltage applied by the audio signal to the varicap alter its capacitance and also the oscillator's frequency. The audio signal amplitude and varicap diode characteristics determine the amount of modulation produced in the circuit.

1.5   Antennas

The final element in any transmission system is the antenna. This part of the system has as its function the transference of energy, in the form of radio waves, into space.

Important characteristics of an antenna are its dimensions and shape (according to the signal frequency and the direction to where the signals must be concentrated). But other factors must be considered to get the best performance from an antenna.

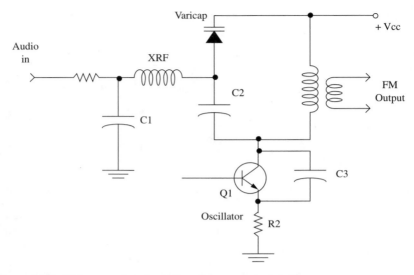

Figure 1.39 Using a varicap to perform frequency modulation.

Figure 1.40 shows the directivity characteristics of some antennas according their shape. The reader can see that some antennas can concentrate all the power from a transmitter in one direction only, whereas others can distribute the energy by spreading the signals in all directions.

The correct antenna is chosen to suit the particular application. If we are using only one receiver to pick up the signals, better performance can be achieved if we concentrate all the transmitter power toward its location. A directional antenna should be used in this case.

A 100:1 concentration factor means that, with the same output power and using a directional antenna, we can transmit to a receiver a signal that is 100 times stronger than if we use an omnidirectional antenna (*omni* is Latin for *all*), which spreads the signals in all directions. Figure 1.41 illustrates. This means that, using a 1 W transmitter, we can get the same performance achieved by a 100 W transmitter under the same conditions.

1.6   Some Practical Circuits

The reader will find more detailed information about transmitter construction in the practical project chapters. However, it is useful at this point to show a few basic circuits.

Circuit 1: 1.5 to 10 MHz Variable Frequency Oscillator

A variable frequency oscillator (VFO) is an oscillator that produces high-frequency, low-power signals in a wide frequency range. The signals are stable and

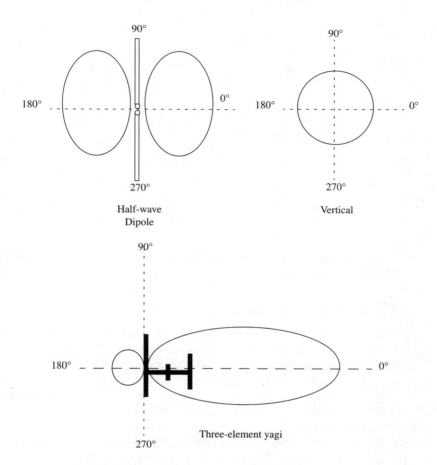

Figure 1.40   Radiation patterns from some antennas.

can be used to drive the power stages of a transmitter. A VFO can be used as the basic oscillator of a transmitter. The circuit shown in Fig. 1.42 produces high-frequency signals covering a wide range of frequencies whose values depend on CV and L1.

Using a coil formed by 60 turns of AWG 28 wire on a ferrite core (1 cm dia. and 10 cm long), the circuit will produce signals in the range of 2 to 4 MHz. By reducing the number of turns to 10 or 12, it is possible to reach the upper frequency limit of approximately 10 MHz.

The signal produced by the first FET is applied to the second FET, where it is amplified again. The second FET acts as a buffer thereby isolating the oscillator stage from the output stage and increasing stability. The capacitors used in the circuit must be appropriate for high-frequency applications such as ceramic or polycarbonate types. The circuit must be powered by a regulated power supply.

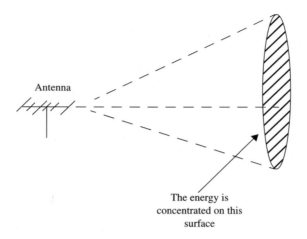

Antenna

The energy is
concentrated on this
surface

Figure 1.41   Directional antennas are used to concentrate the energy of a transmitter
in a small area.

Circuit 2: CW Shortwave Transmitter

Using a 300 Vdc power supply, the circuit shown in Fig. 1.43 can produce a 5 W
signal at some point in the frequency range of 3 to 100 MHz. For the 3 to 8 MHz
band, the coil is formed by 30 turns of AWG 28 or 30 wire on a PVC tube with
diameter of 1 inch (2.5 cm) and a length of 1.5 to 2 inches (3.5 to 5 cm).

The CV must be in the range from 180 to 410 pF or higher capacitances. The
220 and 100 pF capacitors must be rated to 500 WVDC or higher.

The transmitter is mounted on a metallic chassis using a seven-pin socket for
the tube. Lx is the primary winding of a common transformer rated at 117 Vac to
any voltage between 5 and 12 V, with current between 100 and 500 mA. The
electrolytic capacitors must be rated to 300 WVDC or more. The morse key must
be wired between the points A and B in the circuit.

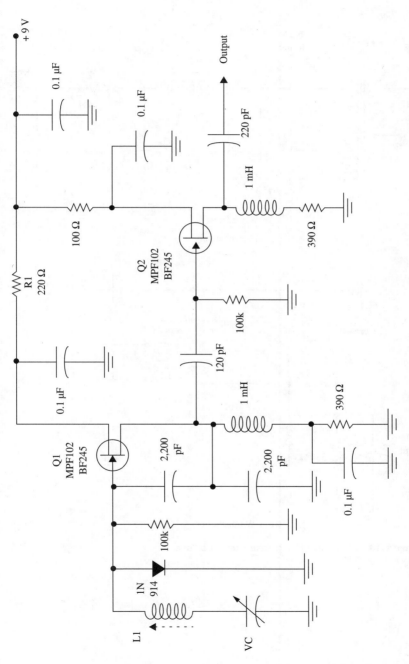

Figure 1.42   1.5 to 10 MHz VFO using FETs.

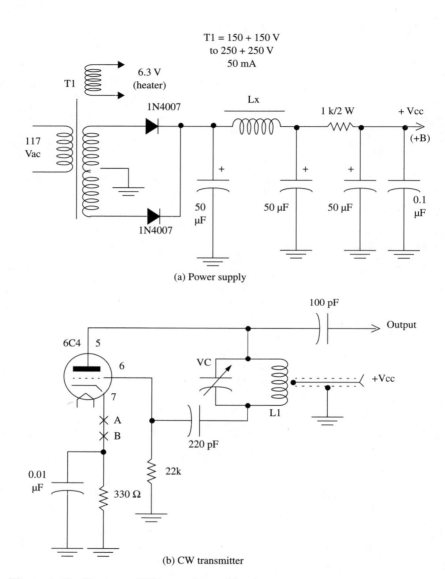

(a) Power supply

(b) CW transmitter

Figure 1.43   Shortwave CW transmitter with tube.

2

FM and VHF Transmitters

This chapter describes FM and VHF transmitters that can be assembled with easy-to-find parts and cheap components. In many cases, components found in old nonfunctioning radios, amplifiers, and TV sets can be used. In less critical cases, the projects can be assembled on either a printed circuit board or a terminal strip. The projects assembled on terminal strips are especially recommended for beginners and students who don't have the practical resources to etch a printed circuit board.

In many cases, the reader can experiment with the project, first mounting it on a terminal strip and later, and if the performance is satisfactory, progressing to a more compact and definitive version using a printed circuit board.

Project 1: Small FM Transmitter

This is the simplest FM transmitter shown in this book. It is an ideal project for the beginner or student.

Features

- Range: 150 to 300 ft
- Power supply voltage: 3 to 6 V (2 or 4 AA cells)
- Frequency range: 88 to 108 MHz (or VHF)
- Number of transistors: 1

This small transmitter can be installed in a small plastic box no larger than a cigarette pack. It can be used as a wireless bug, picking up conversations when hidden in some object near the people who are under surveillance. If coupled with a FM receiver installed in the same box, a pair of these devices can be used as walkie-talkies in bilateral communications.

The signals can travel up to 150 ft (in an open field) if the circuit is powered from 2 AA cells and up to 300 ft if the circuit is powered from 4 AA cells. Any FM receiver can be used to tune the signals at a free point in the 88 to 108 MHz band. It is important to observe that the signal range is affected by solid objects, such as large metallic structures, which can block part of the signal.

Assembly

Figure 2.1 shows the schematic diagram of this simple FM transmitter. The components can be assembled on a small printed circuit board as shown in Fig. 2.2. But, as it is very simple and not critical, it can also be assembled on a terminal strip as Fig. 2.3 shows. Of course, both versions can be installed in a plastic box, which makes for easier transport and additional operational capabilities. We can even recommend this as a first project for a reader who has never before assembled a transmitter or other electronic project.

The coil L is formed by four turns of AWG 22 or 24 wire without a solid core (air core) with a diameter of 1 cm. You can even use a common piece of solid plastic-covered wire for this purpose if AWG enameled wire is not readily available.

The antenna is wired to a tap in the second turn, starting from the collector side of the transistor, and consists in a piece of solid wire 10 to 30 inches long. If a short antenna is used, it can be wired directly to the transistor's collector.

You can also use a telescoping antenna scavenged from any nonfunctioning transistor radio. Don't use larger antennas or external antennas, as these can produce interference in nearly all FM and TV receivers or cause circuit instability.

Any trimmer capacitor with capacitances in the range of 10 to 30 pF can be used for this project. This component is not critical; if the correct frequency is not obtained, you can simply change the number of turns in the coil.

The capacitors must be ceramic types. Many substitutes for the specified transistor can be used. Types such as the BF495, 2N2222, 2N2218, and many others can be used experimentally. Any small RF or high-frequency NPN silicon transistor can be used. You only have to observe the terminal placement if they are not equivalent, as in the BF494.

The electrolytic capacitor must be rated to 6 WVDC or more, and the resistors are 1/8 W or 1/4 W, 5% types. The reader must also observe the position of the battery holder's wires, as this is a polarized component. It is recommended that you do not use a metallic box to house the unit.

Operation

Place near the transmitter 2 to 5 ft away from any FM receiver tuned to a free point in the FM band. Turn on the transmitter and adjust the trimmer until the signal is picked up by the receiver. If you are very close to the receiver, and it is sensitive enough, acoustic feedback will be produced in the form of a loud whistle. Reduce the receiver's volume to stop the feedback. Then, speaking directly into the microphone, your voice will sound loud and clear from the receiver's loudspeaker.

Walk away from the receiver, carrying the transmitter while speaking into the microphone. If the signal disappears as soon as you walk a few feet away, you probably are tuning in a spurious signal rather than the intended frequency. Adjust trimmer again until you find the strongest signal. For the best performance when using the transmitter, stay away from large metallic objects and keep the antenna in a vertical position when speaking. Don't shake the transmitter when you are talking into it.

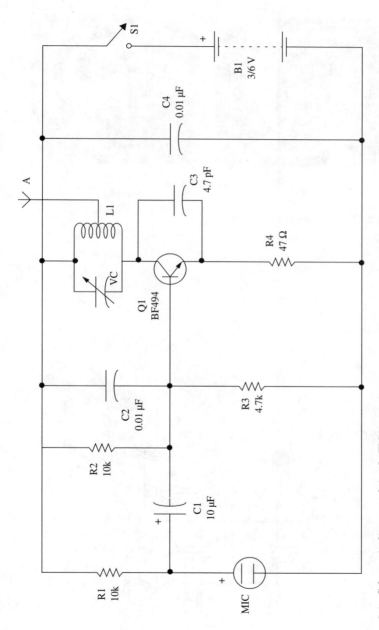

Figure 2.1   Schematic diagram of simple FM transmitter.

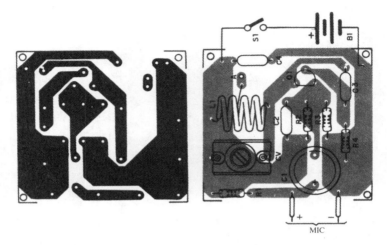

Figure 2.2   Printed circuit board for Project 1.

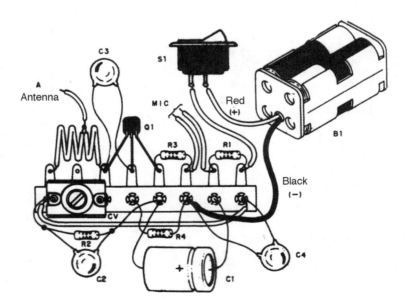

Figure 2.3   Project 1 can also be assembled on a terminal strip.

Parts List: Project 1

Semiconductors

Q1   BF494 or equivalent small signal RF transistor

Resistors (1/8 W, 5%)

R1, R2   10,000 Ω—brown, black, orange

R3   4,700 Ω—yellow, violet, red

R4   47 Ω—yellow, violet, black

Capacitors

C1   10 µF/6 WVDC electrolytic

C2   0.01 µF ceramic

C3   4.7 pF ceramic

C4   0.1 µF ceramic

CV   common trimmer (see text)

Additional Parts and Materials

MIC   electret microphone (two terminals)

L1   coil (see text)

B1   3 or 6 V, 2 or 4 AA cells

S1   SPST toggle or slide miniature switch

Printed circuit board or terminal strip, battery holder, plastic box, solid wire to the antenna, wires, solder, etc.

Project 2: Small FM Transmitter Using a PNP Transistor

A PNP version of the previous transmitter can also be used for short-range communications or as a wireless microphone.

Features

- Power supply voltage: 3 to 6 Vdc
- Frequency range: 88 to 108 MHz (or VHF)
- Range: 150 to 300 ft
- Number of transistors: 1

This transmitter uses a PNP (positive-negative-positive) transistor instead of an NPN (negative-positive-negative) type, which is the basis of most of these projects. Despite this, the circuit has excellent performance as a substitute for those that use NPN transistors.

The charge carriers in an N semiconductor material can flow at higher speeds than in a P semiconductor. As a result, NPN transistors can operate at higher frequencies than equivalent PNP transistors. The NPN transistor has only one P layer to be traversed by the charge carriers, whereas the PNP has two layers. Despite this fact, we can find many PNP transistors that are suitable for use in high-frequency oscillators, reaching frequencies as high as those used in FM transmissions.

In the circuit, we include a PNP Japanese transistor 2SA1177 (or any equivalent) to form a small FM transmitter that resembles the one in Project 1. The recommended transistor has a terminal layout as shown in Fig. 2.4 and can produce signals at frequencies up to 150 MHz.

Using this transistor, we create a small FM transmitter that can send its signals to receivers placed at distances as great as 300 ft, depending on the power supply voltage, the antenna, and the local topographic conditions (presence of hills, obstacles, etc.).

Our project is very interesting from a didactic point of view, as the reader can compare its performance with that of the transmitter shown in Project 1.

The principal characteristics found in the recommended transistor are:

V_{cbo} (max)	30 V
V_{ceo} (max)	20 V
V_{ebo} (max)	5 V
I_c (max)	30 mA
P_c (max)	150 mW
f_T	150 MHz (min); 230 MHz (typical)
h_{FE} (according suffix)	D from 60 to 120
	E from 100 to 200
	F from 160 to 230

These characteristics are provided as a reference for the reader who needs to find an equivalent transistor that is suitable for this application.

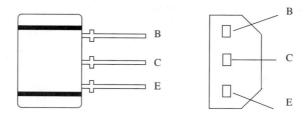

Figure 2.4   Terminal placement for the 2SA1177.

How It Works

The transmitter is formed by a simple one-transistor oscillator that produces the high-frequency signal. The frequency is determined by L1 and CV and can be adjusted to any free point in the FM range between 88 and 108 MHz.

CV is a common trimmer rated 2 to 20 pF, but this component is not critical, and any trimmer with its highest capacitances in the range between 20 and 50 pF can be used.

The feedback that keeps the circuit in oscillation is given by C3. Depending on the operational frequency, this capacitor can be replaced with others to achieve better performance. Values between 2.2 and 5.6 pF can be used experimentally.

Resistor R4 limits the current flow through the transistor and therefore the power. Values below 47 Ω must not be used, as the current flow through the transistor can rise to high values, heating this component to a dangerous point.

If the circuit is powered from a 9 V supply, the signals can reach up to 800 ft, but the value of resistor R4 must be increased to limit the current flow through the transistor to secure values. So, this resistor must be altered to 82 or 100 Ω.

Base biasing is made by resistors R2 and R3, and at the same time C2 decouples this electrode. As this part of the circuit operates with high-frequency signals, all of the capacitors must be ceramic types.

The modulation comes from an electret microphone. Because this kind of microphone has an internal field effect transistor (FET) that is used to increase the signal amplitude (acting as an amplifier), the sensitivity is very good. Of course, the reader can use other microphones such as ceramic or crystal high-impedance types simply by removing R1.

The signals coming from the microphone are applied to the transistor by C1. This capacitor value must be altered if the reader wants to change the audio frequency response that passes through it. Large capacitors will reinforce the bass, and small capacitors will give a good boost to the treble.

The antenna is a piece of solid wire, plastic covered or not, 45 to 150 ft long, such as the one used in Project 1. The same general recommendations about its connection are valid here.

Assembly

The complete schematic diagram of the transmitter is shown in Fig. 2.5. The components are placed on a printed circuit board as shown in Fig. 2.6.

Depending on the power supply (2 or 4 AA cells), the reader can choose an appropriate size plastic box to install all the components. If using AAA cells, it is possible to make the transmitter small enough to be easily transported in your pocket.

The coil consists of 4 turns of AWG 18 to 22 enameled wire with a 1 cm dia. (air core) as in the previous project.

All the capacitors are ceramic types except C1, which could be a plastic film unit or equivalent.

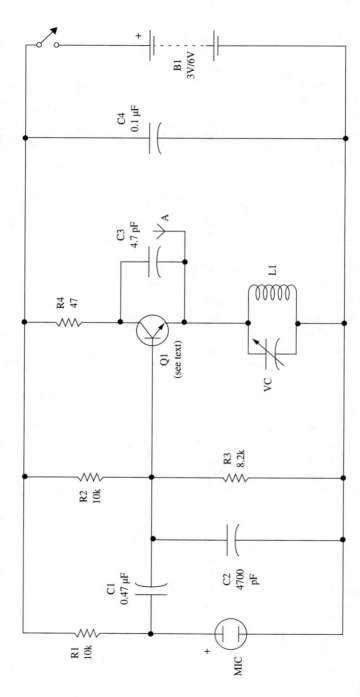

Figure 2.5 FM transmitter using a PNP transistor.

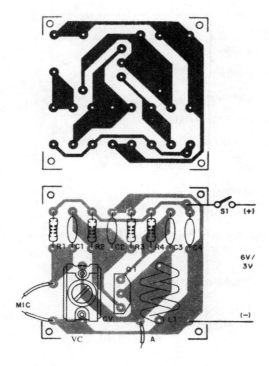

Figure 2.6   Printed circuit board for Project 2.

Adjustment and Use of the Transmitter

The device is adjusted and used as described for Project 1.

Parts List: Project 2

Semiconductor

Q1   2SA1177 PNP RF transistor or equivalent, Sanyo (see text)

Resistors (1/8 W, 5%)

R1, R2   10,000 Ω—brown, black, orange

R3   8,200 Ω—gray, red, red

R4   47 Ω—yellow, violet, black

Capacitors

C1   0.47 μF ceramic or plastic film

C2   0.047 μF ceramic

Parts List: Project 2 (continued)

C3	2.2 to 4.7 pF ceramic
C4	0.1 µF ceramic
CV	trimmer (see text)

Additional Parts and Materials

MIC	electret microphone (two terminals)
S1	SPST toggle, or slide miniature switch
B1	3 to 6 V, 2 or 4 AA cells
L1	coil (see text)

Printed circuit board, battery holder, plastic box, antenna, wires, solder, etc.

Project 3: Five FM Microtransmitters

Using only one printed circuit board, the reader can assemble five different small, low-power FM transmitters.

Features

- Power supply voltages: from 3 to 12 Vdc
- Number of transistor: 1
- Range: 150 ft to 1 mile (depending on the version)
- Frequency range: 88 to 108 MHz

The transmitters described here can send your voice, or any sound picked up by a microphone, to distances between 150 ft and 1 mile, depending on the version. Any FM receiver can be used to tune the signals.

The smallest version can be installed in a matchbox and used as a bug to listen in on secret conversations or simply as a wireless microphone. Figure 2.7 shows some places in a room where a spy transmitter can be hidden.

The highest-power version can send the signals to distances up to 1 mile and can be used for short-range communications or other applications.

Some suggestions for the use of these transmitters are as follows:

- All of the versions can be used as portable transmitters in short-range communications such as wireless microphones. With a VCR camera, you can use the transmitter to send the picked-up sounds to a receiver plugged into the camera's audio input.
- The low-power versions easily can be hidden inside objects to pick up conversations as spy microphones.

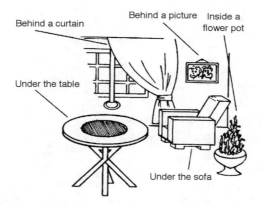

Behind a curtain

Behind a picture   Inside a flower pot

Under the table

Under the sofa

Figure 2.7 Places to hide the spy transmitter.

All versions are simple to assemble and use common parts. You just need the material to etch the printed circuit board and a soldering iron to build all the circuits without encountering any problems.

The principal characteristics of the transmitters are as follows:

Version 1

- Power supply: two button cells (3 V)
- Range: 150 ft (typical)
- Recommended transistor: BF494 or BF495
- Suggested dimensions: 3 × 4 × 1.2 cm

Version 2

- Power supply: two AA cells (3 V)
- Range: 150 ft (typical)
- Recommended transistor: BF494, BF495, or 2N2222
- Suggested dimensions: 3.5 × 4 × 2 cm

Version 3

- Power supply: 2 AA cells (3 V)
- Range: 150 to 300 ft
- Recommended transistor: BF494, BF495, or 2N2222
- Suggested dimensions: 3.5 × 4 × 2 cm

Version 4

- Power supply: 4 AA cells (6 V)
- Range: 600 to 900 ft

- Recommended transistor: BF494 or 2N2218
- Suggested dimensions: $4.5 \times 9 \times 4$ cm

Version 5

- Power supply: battery (9 V)
- Range: 700 to 1,200 ft
- Recommended transistor: BF494 or 2N2218
- Suggested dimensions: $3 \times 8 \times 2$ cm

How the Circuits Work

The small dimensions of these transmitters are the result of their simplicity. In all versions, a single transistor is used as a high-frequency oscillator, producing signals in the range between 88 and 108 MHz. The circuit must be adjusted by the trimmer to operate at a free point in this range.

As the coils are the critical component in this kind of circuit, we have adopted a printed circuit version. The coil is a copper spiral wound on the printed circuit board. It is important to fabricate the printed circuit board carefully, as any interruption or failure of this coil will affect the transmitter's operation.

The feedback that keeps the circuit in oscillation is produced by C4. This capacitor must be a ceramic type with values in the range between 2.2 and 4.7 pF.

Modulation, as in Projects 1 and 2, is produced by a microphone. The signals are applied to the transistor by C1. An electret microphone is recommended, as this kind of transducer has an internal FET to increase sensitivity. Thanks to the FET, it is not necessary to use an external audio amplifier stage to increase the signal's amplitude; they can be used directly to modulate the transmitter stage.

The range of a transmitter such as these depends on several factors, including the power supply voltage, the transistor used, and the antenna size. In each of the different versions, what changes is basically one or more of those factors along with the value of some associated components. For instance, when changing a transistor, it is also necessary to change the bias components wired to it.

From the range of common transistors that can be used to build such a transmitter, we have chosen four types: the BF494 and BF495, which have the same characteristics and can be used in the low-power versions, and the 2N2222 and 2N2218 for the medium- and high-power versions.

The BF494 and BF495 have collector currents rated to 30 mA, although the 2N2218 can operate with collector currents up to 1 A. Of course, this value is not recommended for the applications we have in mind, as the batteries would be drained within a few minutes, and the circuit probably will not oscillate with the expected performance.

It is important to see that, as in the other projects shown in this book, the reader can alter the antenna connection to the transmitter if the transmitter operation is unstable. By connecting the antenna to the proper turn in the coil, it is possible to achieve optimal performance.

Assembly

Printed Circuit Board Etching

Any common method to transfer the original pattern to a virgin copper board can be used by the reader. Photographic methods are the best, and silk screen is suitable if the reader needs more than one unit (for example, to use as a practical project in a technology education class).

The important point to observe is that the copper lines must be uniform, especially in the coil, as any interruption can affect the final circuit performance.

Drilling

A small drill can be used to make holes for the component terminals. A diameter of 0.8 to 1 mm is recommended. If you are using a large drill, it is important to fix the printed circuit board as shown in Fig. 2.8. The wood block placed behind the printed circuit board is important to avoid lateral pressure from the drill, which can rupture the board. Carefully drill holes 1 through 30.

Electronic Assembly

The complete schematic diagram of the transmitter is shown in Fig. 2.9. This diagram is the same for all versions, as component values that change are not specified; instead, they are marked with an asterisk (*). For the correct values, see Table 2.1. Choose your version and look up the component specifications in the table.

The numbers shown in Fig. 2.9 correspond to the points where the components are fixed. For instance, capacitor C2 must have its terminals placed into holes 16 and 17.

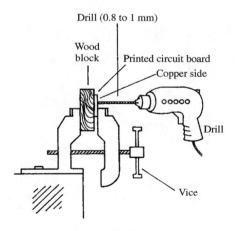

Figure 2.8   Drilling the printed circuit board.

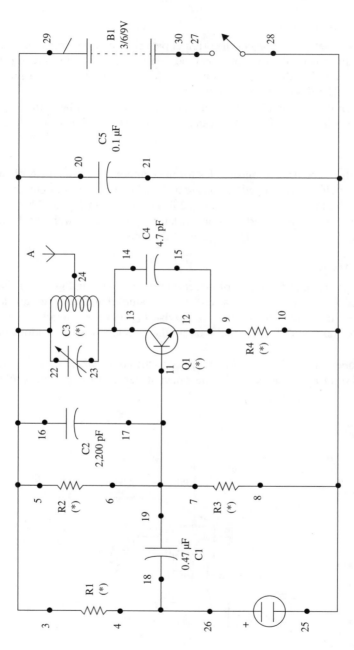

Figure 2.9 Five FM transmitters, schematic diagram for all versions.

Table 2.1 Component Values for Project 3

Version	R1	R2	R3	R4	Q1	Range (ft)
1	1k	3.9k	4.7k	47R	BF494 BF495	up to 150
2	1k	3.9k	4.7k	47R	BF494 BF495	up to 150
	1k	3.9k	4.7k	39R	2N2222	
3	1k	3.9k	4.7k	47R	BF494 BF495	up to 150
	1k	3.3k	4.7k	39R	2N2222	
4	2.2k	5.6k	8.2k	56R	BF494 BF495	up to 300
	2.2k	3.9k	4.7k	39R	2N2218	
5	4.7k	8.2k	10k	120R	BF494	up to 600
	4.7k	4.7k	1.6k	47R	BF495 2N2218	

Color code	
39R	orange, white, black
47R	yellow, violet, black
120R	brown, red, black
1k	brown, black, red
2.2k	red, red, red
3.3k	orange, orange, red
3.9k	orange, white, red
4.7k	yellow, violet, red
5.6k	green, blue, red
8.2k	gray, red, red
10k	brown, black, orange

Passing these terminals through the holes, you have to solder them to the copper side of the printed circuit board. After this operation, cut out the excess terminal wires.

Notice that within the same version the component values can change according to the transistor used. For instance, in version 2, when using a 2N2222 transistor, R2 must be a 3.3 kΩ resistor. If we install a BF494 transistor, we must use a 3.9 kΩ resistor. Be careful to use the correct component values for your version.

In the same table, we show the color codes for all of the recommended resistor values, which will make it easier for the beginner to find these components among stock pieces that may be on hand.

Figure 2.10 shows the terminal placement for the recommended transistors. The resistor can be rated for approximately 1/8 W of dissipation. Larger resistors can be used if the reader can find room for them on the printed circuit board. Figure 2.11 shows the points where the components must be placed.

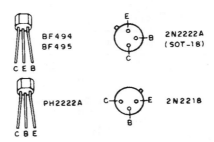

Figure 2.10 Terminal connections for the recommended transistors.

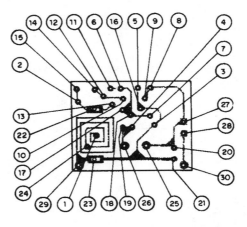

Figure 2.11 Printed circuit board points for component terminal insertion.

Follow the next steps to mount the components on the printed circuit board:

1. *Points 1 and 2.* Between these points we will attach a "jumper" made with a small piece of wire as shown in Fig. 2.12. Cut and bend the wire as shown in the figure and solder it to the printed circuit board's copper side.

2. *Points 3 and 4.* Between these points, solder resistor R1. Note that this component, like many others in the assembly, is placed in a vertical position as shown in Fig. 2.13.

3. *Points 5 and 6.* Install resistor R2. Note that this component's value depends on the transmitter version.

4. *Points 7 and 8.* Resistor R3 is placed between these points.

5. *Points 9 and 10.* The last resistor, R4, is placed between these points.

6. *Points 11, 12, and 13.* Install the transistor at these points. But first see, in Fig. 2.5, the positioning of the terminals for the recommended type. The correct position of each terminal is as follows: point 11, base; point 12, emitter; and point 13, collector.

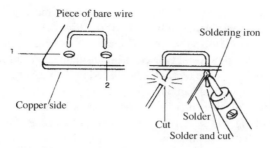

Figure 2.12 Placing the jumper between points 1 and 2.

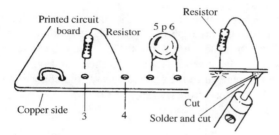

Figure 2.13 Many components must be mounted in a vertical position.

Caution! If any terminal is installed in the wrong hole, the circuit will not operate. Removing the component is not easy and can damage it. Also take care not to touch one terminal to another when installing the transistor on the board. Many hobbyists tend to twist the transistor after it has been soldered, creating a short in the terminals. Don't make this mistake!

---

7. *Points 14 and 15.* These holes are used to install capacitor C4. It is important to use a small disc or plate ceramic capacitor. Take care with the value. It is common to mistake a 4.7 pF capacitor for a 4.7k (4,700 pF) in this application. The circuit will not operate with an incorrect value for this component. Notice that many manufacturers replace the decimal point between the 4.7 pF with a capital letter, indicating the tolerance of the component. A lower-case letter, in contrast, designates the unit. Therefore, 4K7 indicates a 4.7 pF capacitor, but 4k7 means a 4,700 pF capacitor! They are different, and the second one will not work if used as C4 in this project.

8. *Points 16 and 17.* C2 is placed between these points. Also take care with the value of this component. Labels such as 222, 22k, 2,200, or 2n2 are common.

9. *Points 18 and 19.* Place capacitor C1 between these points and solder its terminals to the copper side of the board. The values may be marked as 474, 470k, .47, or 0.47.

10. *Points 20 and 21.* Capacitor C5 is placed between these points. The values marked on this component can be 104, 100k, .1, or 0.1.
11. *Points 22, 23, and 24.* The trimmer C3 and the antenna are the next components to be installed. Several types of trimmers can be used for this project. It is important only that the terminals be placed in the proper holes.

    In our project, we used an old porcelain-type trimmer, but plastic types are suitable. Before placing the trimmer on the board, cut a piece of rigid wire 5 to 25 inches long and solder it in hole number 24. It is the antenna. Afterward, you can fix the trimmer as shown in Fig. 2.14. If the trimmer terminals are larger in diameter than the holes, you can solder a small piece of rigid wire to each and use those wires as terminal extensions, passing them through the holes.
12. *Points 25 and 26.* We complete the work on the printed circuit board by soldering the wires to the microphone. Notice that the microphone is a polarized component, which means that you must observe the existence of a positive terminal (+) and a negative terminal (−) as shown in Fig. 2.15. Each of those poles must be connected to the correct wire. If the connections are inverted, the circuit will not work properly.

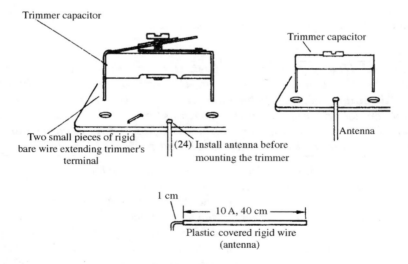

Figure 2.14   Installing the antenna and trimmer capacitor.

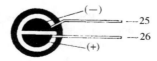

Figure 2.15   Wiring the microphone.

13. *Points 27 and 28.* Switch S1 is installed here. The printed circuit board was designed to use a small miniature SPST switch as shown in Fig. 2.16. This switch can be prepared as shown in the figure before being placed on the circuit board. But you can also use any other SPST switch if you wire it as shown in the figure, using two pieces of common wire. In this way, it is possible to install the switch directly in the plastic box that houses the circuit.

It is also acceptable to eliminate this component. You can turn off the transmitter simply by removing the cells from the battery holder.

14. *Points 29 and 30.* These holes are used for the battery holder connection. Notice that the holder is also a polarized component. The proper position of the red (+) and black (–) wires must be observed. The red wire is connected to point 29 and the black to point 30.

If using a 9 V battery, the reader must use an appropriate connector or clip and also observe the position of the wires. And, when using two button cells as shown in Fig. 2.17, the reader must install the holder. Two wires are placed in contact with the batteries' electrodes and all the parts are kept together by the use of an adhesive band.

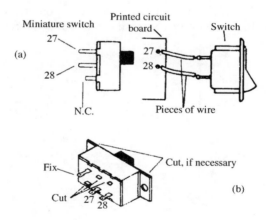

Figure 2.16   Installing S1.

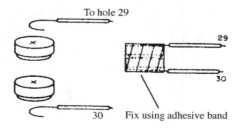

Figure 2.17   Using button cells to power the transmitter.

Testing and Using the Transmitter

Place any FM receiver near the transmitter (3 to 6 ft away) and tune it to a free point in the FM band. Set the volume to 1/3 to 1/2 of the maximum.

Put the transmitter's batteries in the battery holder and turn on S1. Using a plastic or wood tool, adjust the trimmer. (Wood and plastic tools for adjusting coils and trimmers can be found in electronics specialty shops.)

The signal can be tuned quickly and will be heard as a strong whistle. The whistle is caused by the acoustic feedback, which can be eliminated by cutting back on the receiver's volume.

You can now speak into the microphone to test the modulation. As you walk a greater distance from the receiver, the signal must be more finely tuned. If the signal disappears when you have moved a few feet away, you are certainly tuned to a spurious signal or harmonic rather than the main frequency. Make further tuner adjustments to find a stronger signal. After you have properly tuned the transmitter, you can also determine the correct length for the antenna experimentally. After your transmitter is working, you can install all the pieces in a plastic box for physical and electrical protection as shown in Fig. 2.18.

Remember that you will obtain the best results when using your transmitter in open field, because large metal structures can act as obstacles to the signals' propagation. Cell life depends on the transmitter power and other variations from one circuit to another.

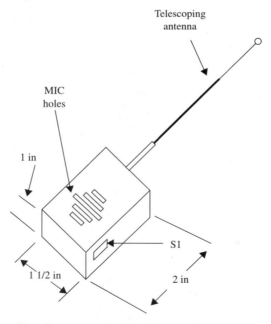

Figure 2.18 Installing the transmitter in a plastic box.

Parts List: Project 3

Semiconductor

Q1   BF494, BF495, 2N2222, or 2N2218, according to the version, NPN
                 silicon RF transistor

Resistors (1/8 W, 5%)

R1, R2, R3, R4   (see table; according to the version)

Capacitors

C1   0.47 µF, any type

C2   2,200 pF ceramic

C3   trimmer (see text), (2–20 to 4–40 pF), plastic or porcelain

C4   4.7 pF ceramic

C5   0.1 µF ceramic

Additional Parts and Materials

MIC   electret microphone (two terminals)

B1   3 to 9 V (according to the version) cells, battery, etc.

L1   coil, printed on the circuit board

S1   SPST miniature switch (see text)

Printed circuit board, antenna, battery holder, wires, solder, etc.

Project 4: High-Power FM Transmitter

This transmitter is powered from a 12 V supply and can send signals in the FM
band up to 1 mile under favorable conditions.

Features

- Power supply voltage: 12 V (battery or power supply)
- Frequency range: 88 to 108 MHz
- Range: more than half a mile
- Microphone: low impedance

High-power FM transmitters are very popular among many readers, although
regulatory restrictions on their operation limit their practical use. However, a
high-power transmitter such as described here can be used on farms, campsites,
and other large properties located far away from cities where its operation might
interfere with TV and other telecommunication equipment.

The reader who wants a powerful transmitter for experiments or practical applications (within limits) can build this interesting simple circuit using only one transistor. The original version uses a 2N2218 transistor to achieve a broadcast range of up to a mile, but it is also possible use a BD135 transistor. Although originally designed for audio applications, this component can oscillate in the FM range to achieve the same performance.

Figure 2.19 shows how the circuit can be installed into a small box to create a portable version. This circuit uses a rechargeable 12 V nicad battery or, if the reader prefers, 6 to 8 AA alkaline cells. Of course, the life span of AA cells is not as long that of the NICAD battery.

How It Works

The high-frequency oscillator runs at 88 to 108 MHz, putting the signal at a free point in the FM range. This oscillator is of the same basic configuration using one transistor 2N2218 as shown in previously described projects.

The signal frequency is determined by the resonant circuit formed by L1 and the trimmer capacitor CV. The trimmer must be adjusted to tune the circuit to a free point in the FM band. The high-frequency signal produced by this oscillator is applied to the antenna by C4.

Once again, we recommend that the reader avoid instabilities that may result from wiring the antenna directly to the transistor. You should connect it to an appropriate tap in the coil. By matching the antenna impedance with the circuit impedance, it is possible to get better performance, i.e., avoiding instability and increasing the output power. Figure 2.20 shows how the antenna can be connected to the output stage.

The modulation comes from an electret microphone, as in previously described projects, or from a small loudspeaker. Figure 2.21 shows how an electret microphone can be wired to this transmitter.

Finally, it is possible to wire the output of a CD player or tape deck into this circuit to broadcast music. Therefore, the project can be used as an experimental radio station.

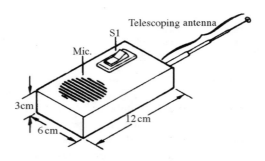

Figure 2.19   Installing the transmitter in a small box.

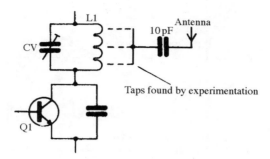

Figure 2.20 Connecting the antenna to L1.

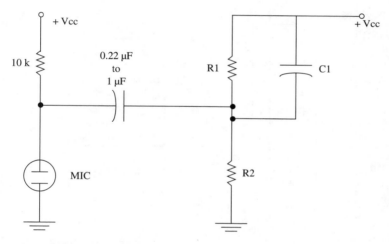

Figure 2.21 Using an electret microphone.

Assembly

Figure 2.22 shows the complete schematic diagram of the transmitter. Because the circuit is not critical, if the reader keeps the terminals and wires short, a terminal strip can be used as the chassis as shown in Fig. 2.23. But remember that, if a terminal strip is used, a larger box will be necessary to house the circuit. Alternatively, the components can be placed on a printed circuit board as shown in Fig. 2.24.

The coil is formed by 4 or 5 turns of AWG 22 or 24 enameled wire in a 1 cm diameter form without an iron core (i.e., air core). Taps in the second or third turn can be tested for a proper antenna connection.

Any transistor audio output transformer with primary impedances rated to values between 200 and 2,000 Ω and an 8 Ω secondary winding can be used to connect the low-impedance microphone.

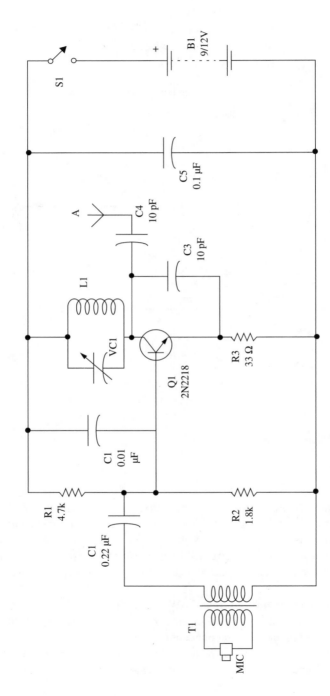

Figure 2.22 High-power FM transmitter.

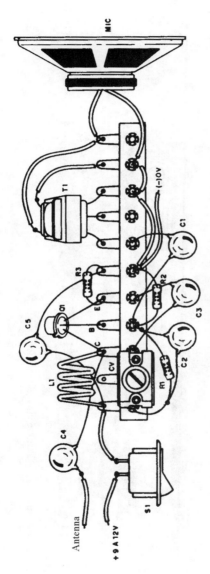

Figure 2.23   A terminal strip can be used as the chassis for this project.

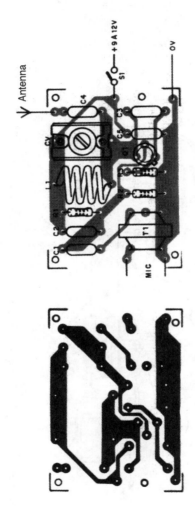

Figure 2.24 Printed circuit board for Project 4.

The microphone is a small transistor radio loudspeaker or any low-impedance unit (4 to 8 Ω). The antenna is a piece of solid wire (plastic covered) or a telescoping antenna as used in transistor radios with lengths in the range of 6 to 30 inches.

Adjustment and Operation

Do not power the circuit from a common 9 V battery. As the current drain is high, such a battery would fail in a few minutes.

The procedure to tune the circuit is the same as described in the previous projects: place an FM receiver close to the transmitter and look for a free point in the dial. Adjust CV to produce the strongest audio signal from the receiver. Speak into the microphone to test the modulation. If any acoustic feedback is noted, reduce the receiver volume or move it farther away from the transmitter.

Parts List: Project 4

Semiconductor

Q1   2N2218 or BD135 NPN transistor (see text)

Resistors (1/8 W, 5%)

R1   4,700 Ω—yellow, violet, red

R2   2,200 Ω—red, red, red

R3   39 Ω—orange, white, black

Capacitors

C1   0.22 µF ceramic

C2   0.01 µF ceramic

C3, C4   10 pF ceramic

C5   0.1 µF ceramic

CV   trimmer (see text)

Additional Parts and Materials

T1   audio output transformer, 1,000 Ω × 8 Ω (see text)

L1   coil (see text)

MIC   8 Ω × 1 to 2 inches, small loudspeaker

S1   SPST toggle or slide switch

B1   9 to 12 V, AA cells or nicad battery

Printed circuit board or terminal strip, plastic box, antenna, battery holder, wires, solder, etc.

Project 5: FM Transmitter Using an IC

This circuit uses an operational amplifier (opamp) integrated circuit to increase the microphone's signal gain, increasing sensitivity. The signal can be broadcast to distances up to a half a mile using a 9 V power supply.

Features

- Power supply voltage: 6, 9, or 12 Vdc
- Output power: 200 mW (typical)
- Frequency range: 88 to 108 MHz (or VHF)
- Range: half a mile and more
- Current drain: from 50 mA to 100 mA (typical)

Here we describe a small FM (or VHF) transmitter intended for general purposes and also for espionage. The transmitter can send the signals to distances up to 1,500 ft if powered from a 6 V supply, and 1 mile if powered from a 9 V supply.

The basic difference between this circuit and the previously described ones is the presence of an opamp in the modulation stage. With a very high gain characteristic, the circuit can increase the amplitude of very weak sounds picked up by the microphone before applying the audio signals to the transmitter. This means that the circuit can be used for remote listening, as required in espionage, where the sounds to be heard are very weak.

Another important point to observe is that the amplifier gain can be programmed by only one component. This circuit, as many others described in this book, can also be installed in a small plastic box as shown in Fig. 2.25.

How It Works

The high-frequency stage has the conventional configuration seen in the projects presented earlier. The transistor can be a 2N2218, 2N2219, or a BD135, all of

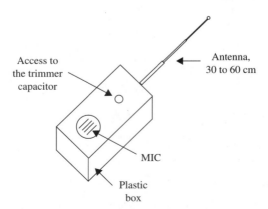

Figure 2.25 The transmitter can be installed in a plastic box as shown.

which are very easy to find. These transistors have collector currents rated up to 1 A and can produce a strong signal in the FM range as well as in the low VHF band between 40 and 88 MHz.

The operational frequency is determined by the resonant circuit L1/CV. CV must be adjusted to a free point in the FM band as recommended in the other projects. It is also possible to change the coil to tune the circuit to the low VHF band. Of course, the reader must have a receiver that can receive that band.

The feedback that maintains the oscillation is given by C5. This capacitor must be a ceramic type for best performance.

The transistor base is polarized by R6 and R7, and the current between collector and emitter is limited by R8. By increasing the value of R8, it is possible to reduce power consumption, but the output power will be also reduced. The advantage is that this procedure extends battery life.

Modulation is provided by an amplifier stage based in a single opamp. If the circuit is powered from a 6 V supply, the opamp must be a low-voltage type such as the CA3140. But if the circuit is powered from a 9 V supply, you can use the 741. The 741 IC is a very common component and can be obtained easily from many dealers. This IC's non-inverting input is wired to a voltage divider formed by R3 and R4. This divider gives half of the supply voltage to bias this input.

The audio signal is applied to the inverting input, where we also find the feedback resistor R5. This component fixes the amplifier gain. The reader can make experiments with this component, trying to find a value appropriate for the amplitude of the sounds he intends to pick up. For strong sounds, for instance, a resistor between 0 (i.e., a short circuit between pins 2 and 6) and 100 k can be used. For weak sounds, the resistor must be between 150 kΩ and 1.5 MΩ. The suggested value of 220 kΩ is indicated for both weak and medium-intensity sounds.

The radio signals are sent into the space via a telescoping antenna or a piece or wire 20 to 40 inches long. As with other projects, the reader can determine the proper tap in the coil for the antenna connection.

Assembly

The circuit of this transmitter is shown in Fig. 2.26. The circuit must be assembled on a printed circuit board as shown in Fig. 2.27. Figure 2.28 shows how a small heat sink can be placed on the transistor.

If you find it difficult to work with ICs, you may find it easier to install that device in an eight-pin dual in-line package (DIP) socket.

The coil L1 consists of 4 turns of AWG 18 to 22 enameled wire on a coreless form 1 cm in diameter. The tap for the antenna will be at the second or third turn. To operate the circuit in the range between 40 and 80 MHz, the coil will be formed by 6 to 8 turns of the same wire in a 1 cm diameter coreless form (air core).

Any plastic or porcelain trimmer with capacitances in the range between 20 and 40 pF can be used (2–20 to 4–40).

An interesting alternative to the LC circuit is optional. Use a common intermediate frequency (IF) transformer coil (with ferrite core) and fix the parallel

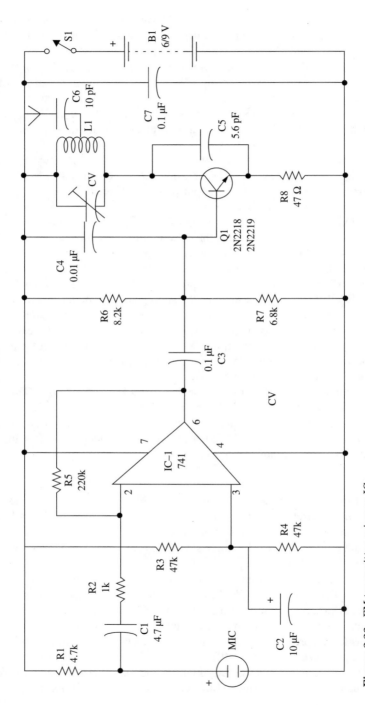

Figure 2.26 FM transmitter using an IC.

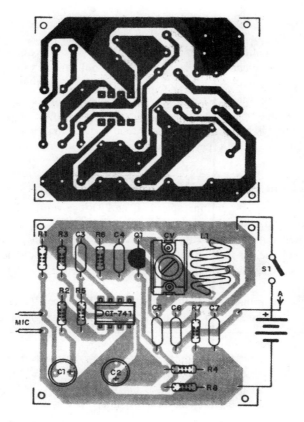

Figure 2.27   Printed circuit board for Project 5.

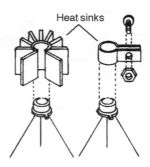

Figure 2.28   Placing the heat sink on Q1.

ceramic capacitor (replacing CV) at 2.2 or 3.3 pF. This way, you can tune the circuit using the ferrite core instead of the trimmer.

Tolerance, dissipation, and voltage ratings for all the components are indicated in the parts list. For electrolytic capacitors, the recommended values are lower.

The electret microphone can be placed far from the circuit but, in this case, it must be wired using shielded cable to avoid noise and hum.

The recommended power supply is formed by six to eight AA or D cells. Another power supply choice is a nicad cell. Don't use a small 9 V battery for this task, as the current drain is very high, and the battery will run down in just a few minutes.

Testing and Using the Circuit

The procedure is the conventional one: tune an FM receiver to a free point in the FM (or VHF) band. Place the receiver near the transmitter.

Tune the transmitter by adjusting CV. If acoustic feedback is produced, reduce the receiver volume. As the circuit is very sensitive to weak sounds, it would be better to use an earphone plugged into the receiver when making these adjustments.

How to use the transmitter is up to the reader. For surveillance applications, place it far away from large metallic objects that can affect signal propagation.

Parts List: Project 5

Semiconductors

IC1	741 operational amplifier, integrated circuit (or CA3140 if the circuit is powered from a 6 V supply)
Q1	2N2218 or 2N2219 NPN transistor (see text)

Resistors (1/8 W, 5%)

R1	4,700 Ω—yellow, violet, red
R2, R3, R4, R8	47,000 Ω—yellow, violet, orange
R5	220,000 Ω—red, red, yellow
R6	8,200 Ω—gray, red, red
R7	6,800 Ω—blue, gray, red
R8	47 Ω—yellow, violet, black

Capacitors

C1	4,7 µF/12 WVDC, electrolytic
C2	10 µF/12 WVDC, electrolytic
C3, C7	0.1 µF ceramic

Parts List: Project 5 (continued)	
C4	0.01 µF ceramic
C5	5.6 pF ceramic
CV	trimmer (see text)

Additional Parts and Materials

MIC	electret microphone (two terminals)
S1	SPST toggle or slide switch
L1	coil
B1	6 to 12 V, battery, cells, etc. (see text)

Printed circuit board, plastic box, battery holder, telescoping antenna (optional), wires, solder, etc.

Project 6: Two-Transistor FM Transmitter

This small FM transmitter can send the signals to a receiver placed at distances up to 1,500 ft.

Features

- Power supply voltage: 6 to 12 Vdc
- Frequency range: 88 to 108 MHz
- Number of transistors: 2
- Range: 1,500 ft (typical)

This transmitter can be powered from a 6 to 12 V power supply, dry cells, or a nicad battery. The range will depend on the supply voltage. The circuit also implements an audio amplifier stage to increase microphone sensitivity.

Observe that only one transistor is used to produce the RF or high-frequency signals. The other transistor is used for the audio signals.

The reader can use this transmitter as a wireless microphone for short-range communications or as a "bug" to hear conversations in an adjacent room.

The ideal bias for the audio stage can be adjusted using a trimmer potentiometer, allowing the operator to achieve the best performance. Best results are obtained when using a 3 to 4 ft antenna. Figure 2.29 shows the complete schematic diagram for the transmitter. The components with values enclosed in parentheses are indicated for a 12 V power supply.

L1 and CV are the same as used in previous projects. All capacitors with values less than 1 µF must be ceramic types. The higher-value components are electrolytic, rated to voltages as specified in the parts list.

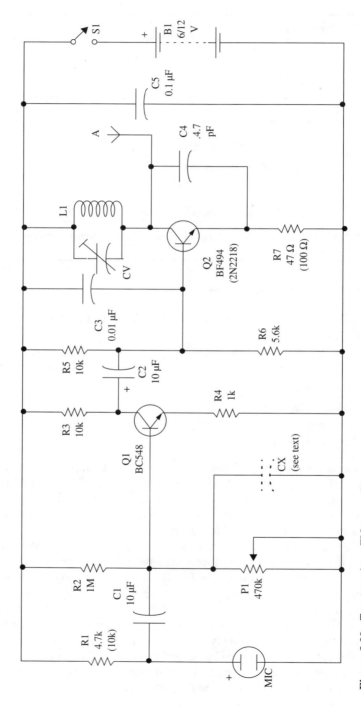

Figure 2.29 Two-transistor FM transmitter.

The electret microphone can be replaced with a common high-impedance microphone if you remove resistor R1. If you want to increase the sensitivity to low-frequency sounds (bass) by cutting the high frequencies, you can insert a capacitor (CX) as shown in the diagram. The proper value of this component is found experimentally, but it probably will be in the range of 0.1 to 10 μF. The components are placed on a printed circuit board as shown in Fig. 2.30.

To adjust and use this transmitter, proceed as described for the previous projects.

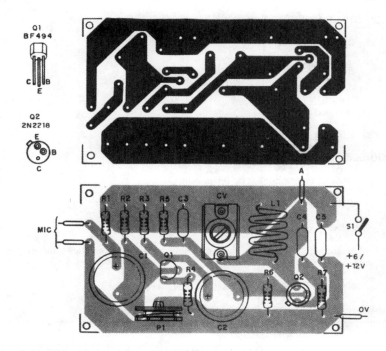

Figure 2.30   Printed circuit board for Project 6.

Parts List: Project 6

Semiconductors

Q1   BC548 or equivalent general-purpose NPN silicon transistor

Q2   BF494 or equivalent RF silicon NPN transistor

Resistors (1/8 W, 5%)

R1   4,700 Ω—yellow, violet, red (10,000 Ω—brown, black, orange)

R2   1,000,000 Ω—brown, black, green

Parts List: Project 6 (continued)

R3, R5   10,000 Ω—brown, black, orange

R4   1,000 Ω—brown, black, red

R6   5,600 Ω—green, blue, red

R7   47 Ω—yellow, violet, black (100 Ω—brown, black, brown)

Capacitors

C1, C2   10 μF/16 WVDC, electrolytic

C3   0.01 μF ceramic

C4   4.7 pF ceramic

C5   0.1 μF ceramic

Cx   (see text)

CV   trimmer (see text)

Additional Parts and Materials

MIC   electret microphone (two terminals)

L1   coil (see text)

S1   SPST toggle or slide switch

A   telescoping antenna or a piece of rigid wire 10 to 20 inches long (plastic covered)

B1   6 to 12 V, AA cells, nicad battery, power supply, etc.

Printed circuit board, battery holder, plastic box, wires, etc.

Project 7: Two-Stage FM Transmitter

This high-power transmitter uses one transistor to produce the high-frequency signals and a second transistor to power an amplifier stage. Using this configuration, the range is up to 2 miles in open field.

Features

- Power supply voltage: 6 to 12 Vdc
- Output power: 200 mW to 1 W
- Frequency range: 88 to 108 MHz
- Number of transistors: 2
- Range: up to a few miles

This transmitter uses a power output stage to increase the high-frequency signal power. The output power rises to a few hundred milliwatts, which is enough to send signals to a receiver placed at distances up to two miles.

The power supply used in this circuit must produce a large amount of current. Therefore, it is recommended that you avoid the use of common cells in favor of supplies powered from the ac power line, a car battery, or nicad battery. If the circuit is powered from the ac power line, it is very important to use good filtering and regulation to avoid noise and hum in the transmitted signals.

The main application suggested for this transmitter is as an experimental broadcast station for clubs, schools, or campgrounds. The signals from a CD player, tape deck, or microphone passing through a mixer can be applied to the transmitter's input to act as program sources.

The transistor is not an expensive type as would be recommended for high-power transmitters; rather, it is a common, low-cost audio transistor that can oscillate in the FM range. The transistor's low cost is an additional attraction for the reader who doesn't have much money to spend on a transmitter project. Only two adjustments are needed to put this transmitter on the air, and building it does not require any special tools or instruments.

How It Works

The oscillator stage is formed by a BF494 or BF495 in the configuration seen in previous projects. This stage produces a signal with a frequency determined by L1 and CV. The reader must tune this circuit to a free point in the FM band between 88 and 108 MHz.

The feedback that keeps the oscillator on is given by C3, the capacitor wired between the transistor's collector and emitter. Resistors R1 and R2 bias the transistor's base, and C2 decouples this electrode, applying high-frequency signals to ground (passing through the power supply). The audio signal is applied to the circuit via C1 as in many other circuits described in this book.

The audio signal's amplitude must be controlled to avoid saturating the circuit, causing distortion. Audio signal sources can be used, such as a mixer or sound console processing signals from sources such as tape decks, CD players, microphones, etc.

The signal produced by this stage is applied to a power output stage using a BD135, BD137, or BD139. The BDs are audio, medium-power transistors, but they have a good gain when working with high-frequency signals. Their f_T (transition frequency) is about 150 MHz, which means that they can produce a good power gain when working with signals in the FM band.

To get the best performance from this transistor, the configuration must be the so called "common base" one as shown in Fig. 2.31. The base/emitter capacitance, which decreases the high-frequency gain, can be reduced if the transistor is wired in this configuration. Of course, the power gain found in this configuration is not as high as in the common emitter configuration, but the capacitance effects can be reduced.

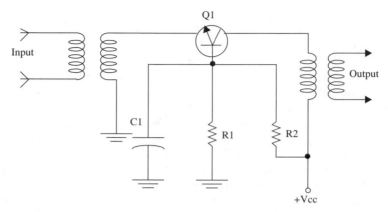

Figure 2.31 Common base stage used in transmitters.

Figure 2.32 shows what happens. Large gains at higher frequencies can be achieved if the transistor is used in the common base configuration.

The high-frequency modulated signal is applied to the amplifier stage by L2. Observe that, to get the better performance when transferring the high-frequency signal from one stage to another, it is necessary to adjust CV1 correctly.

Resistors R5 and R6 bias the output transistor, and C5 decouples its base. The output signal appears on CV2/L3 and, from this resonant circuit, it can be transferred to the antenna.

The other necessary adjustment is to tune CV2 for best performance in signal transference. Capacitors C6 and C7 filter and decouple the power supply voltage.

Assembly

The complete diagram of this transmitter is shown in Fig. 2.33. The printed circuit board suggested for this project is shown in Fig. 3.34.

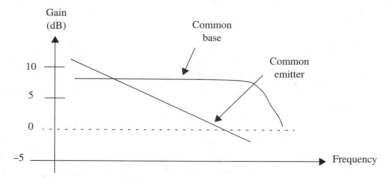

Figure 2.32 Common base stages operate better in high-frequency stages than in common emitter stages.

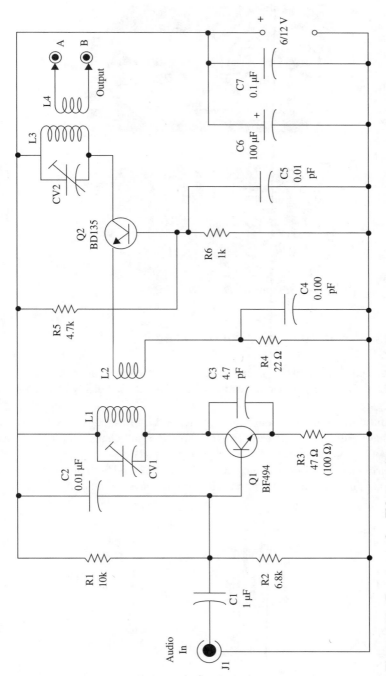

Figure 2.33 Two stages of an FM transmitter.

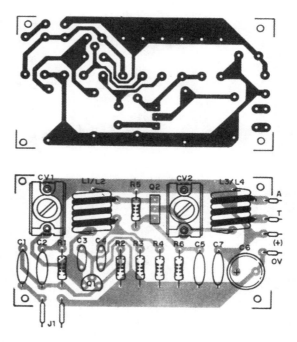

Figure 2.34 Printed circuit board for Project 7.

L1/L2 and L3/L4 must be assembled one inside the other (interlaced) as shown in Fig. 2.35. The coils are wound using AWG 18 to 22 enameled wire as shown in same figure.

L1 and L3 is formed by four turns, and L2 and L4 formed by three turns. The forms have a diameter between 0.6 and 0.8 mm. Small changes in these dimensions can be compensated by the trimmer adjustments.

The trimmer capacitors can range from 20 to 40 pF, and both plastic and porcelain types can be used. According to the terminal layout found in these compo-

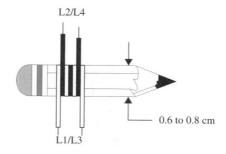

Figure 2.35 Using a pencil or a tube as a form to wire the coils.

nents, the need to make small changes in the printed circuit layout must be considered. We recommend that the reader first buy the trimmer capacitors and see how their terminals are placed. Afterward, you can work on the printed circuit board project and make etching alterations if necessary.

All capacitors except C6 are ceramic types (disc or plate). The electrolytic is rated to 16 WVDC or more.

The transistor can be a BD135, BD137, or BD139, and it must be mounted on a small heat sink. It is interesting to experiment with several transistors to achieve the best gain. These transistors have a wide range of characteristics, and among many apparently identical units it is possible to find one with higher gain.

For the audio signal input, you can use a jack that is appropriate for the signal source. If an electret microphone is used, a 10 k Ω resistor must be used for biasing as shown in Fig. 2.36.

The component values in parentheses (Fig. 2.33) are appropriate when using a 12 V power supply. A regulated power supply for this circuit is shown in Fig. 2.37. The transformer has a primary winding rated to 117 Vac and a secondary rated to 12 V (CT) × 1 A.

The electrolytic capacitors must be rated to 25 WVDC or more. The power-on LED indicator is optional.

Adjustments and Use

Tune any FM receiver to a free point of the FM range and place it 10 to 15 ft away from the transmitter. To the transmitter's output, wire a small 6 V × 50 mA incandescent lamp or a multimeter adjusted to the 0–5 Vdc scale as shown in Fig. 2.38. First, tune CV1 until you can hear the signal in the receiver. Second, adjust CV2 to the highest output power (higher lamp glow or higher voltage

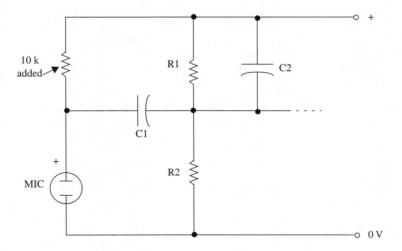

Figure 2.36   Using an electret microphone.

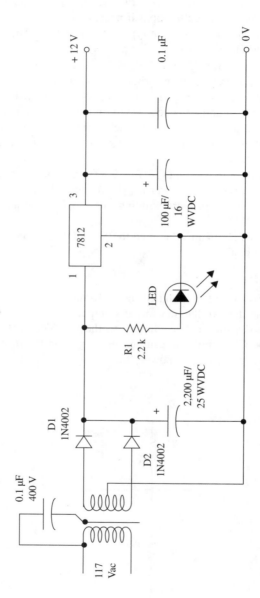

Figure 2.37 Power supply intended for this transmitter.

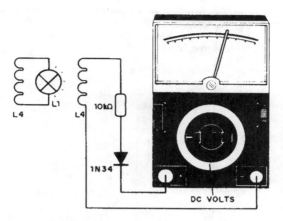

Figure 2.38 Using a multimeter to make adjustments.

indicated by the multimeter). Readjust CV1 adjustments to keep the signal in the tuned point in the receiver.

An external antenna, if installed, can be connected to the transmitter as shown in Fig. 2.39. Using a field strength meter, adjust CV3 to get the best performance. An FM receiver with an S-meter can be used for this purpose. A shielded cable must be used to connect the transmitter to the antenna.

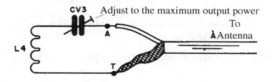

Figure 2.39 Connecting the transmitter to an external antenna.

Parts List: Project 7

Semiconductors

| Q1 | BF494 or BF495, NPN high-frequency (RF) transistors |
| Q2 | BD135, BD137, or BD139, medium-power NPN silicon transistors |

Resistors (1/8 W, 5%)

R1	10 k Ω—brown, black, orange
R2	6,800 Ω—blue, gray, red
R3	47 Ω—yellow, violet, black (100 Ω × 1 W, brown, black, brown)

Parts List: Project 7 (continued)

R4	22 Ω—red, red, black
R5	4,700 Ω—yellow, violet, red
R6	2,200 Ω—red, red, red
Capacitors	
C1	1 μF metal film or ceramic
C2, C5	0.01 μF ceramic
C3	4.7 pF ceramic
C4	100 pF ceramic
C6	100 μF/16 WVDC electrolytic
C7	0.1 μF ceramic
CV1, CV2	trimmer (see text)
Additional Parts and Materials	
J1	microphone jack
L1, L2, L3, L4	coils (see text)

Printed circuit board, power supply, plastic box, antenna terminals, heat sink to Q2, wires, solder, etc.

Project 8: Three-Transistor FM Transmitter

This large range transmitter can send signals to distances up to a mile and even more depending on the power supply voltage and some other local factors as described in chapter one.

Features

- Power supply voltage: up to 12 Vdc
- Frequency range: 88 to 108 MHz
- Range: several miles
- Number of transistors: 3

This transmitter uses three transistors. Two are used to form two high-frequency stages, and the third is used as an audio amplifier in the modulation stage.

The circuit needs three adjustments. Two are made using trimmer capacitors that act on the high-frequency signals to achieve maximum power output, and the other is made with a trimmer potentiometer. The trimmer potentiometer adjusts the audio input stage to obtain better performance according to the magnitude of the modulation signal.

An external antenna can be used under the conditions described in Chapter 1. With an external dipole, the signals can be sent distances up to several miles over an open field.

For experimental purposes, we recommend the use of a telescoping antenna with ranges between 15 and 40 inches or a piece of wire with the same length.

Figure 2.40 shows the schematic diagram of this transmitter. The components placement on a printed circuit board is shown in Fig. 2.41.

All of the coils can be wound on a pencil using AWG 18 to 22 enameled wire (or even plastic-covered wire). The coils must have the following number of turns:

L1	4 turns
L2	3 turns interlaced with L1
L3	4 turns
L4	5 turns

Any plastic or porcelain trimmer capacitor with capacitances in the range of 2–20 to 4–40 pF can be used. The differences between these trimmer capacitors and other component tolerances can be compensated by changing the number of turns in L1 and L3.

All capacitors except C1 and C2 are ceramic disc or plate types. C1 and C2 are electrolytic capacitors rated to 16 WVDC or more.

The circuit can be powered from supplies ranging from 6 to 12 V. If a 9 to 12 V supply is used, Q3 must be mounted on a heat sink. Equivalents of transistor Q3 are the BD137 and BD139.

Adjustments are made as described for the transmitter shown in Project 7. The difference is that, after adjusting CV1 and CV2 to get the best performance (high signal output), you must also adjust CV3 to match the output impedance with the antenna impedance to broadcast a strong signal into space. Use a field strength meter to make this adjustment. When applying an audio signal to the input, you also must adjust P1 to find the best audio reproduction, avoiding distortion due overmodulation.

For operation between 50 and 80 MHz, change the coils to the following:

L1	5 or 6 turns
L2	4 turns interlaced with L1
L2	5 or 6 turns
L3	8 turns

The wires and the form remain the same. The only component change is to C4, which must be increased to 10 pF.

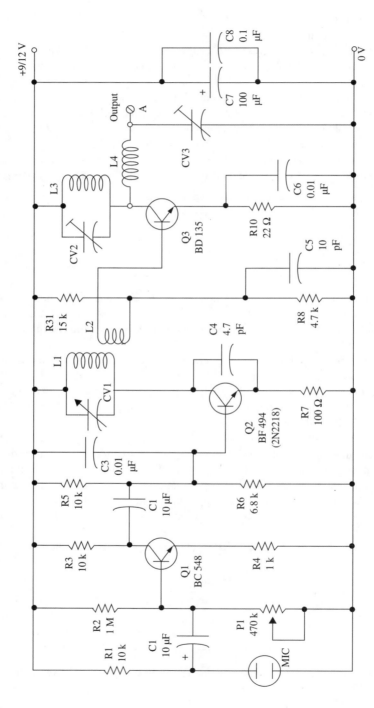

Figure 2.40 Three-transistor transmitter.

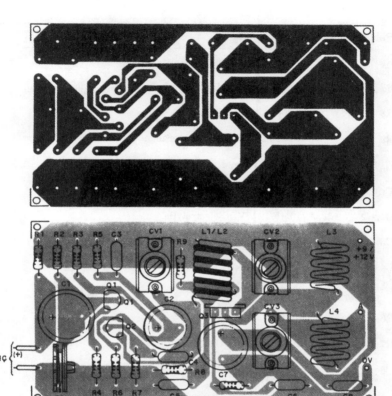

Figure 2.41   Printed circuit board for Project 8.

Parts List: Project 8

Semiconductors

Q1	BC548 or equivalent general-purpose NPN silicon transistor
Q2	BF494 (6 V) or 2N2218 (12 V) - NPN silicon transistor
Q3	BD135 or equivalent medium-power NPN silicon transistor

Resistors (1/8 W, 5%)

R1, R3	10,000 Ω—brown, black, orange
R2	1,000,000 Ω—brown, black, green
R4	1,000 Ω—brown, black, red
R5	10,000 Ω—brown, black, orange

Parts List: Project 8 (continued)

R6	6,800 Ω—blue, gray, red
R7	100 Ω—brown, black, brown
R8	2,200 Ω—red, red, red
R9	15,000 Ω—brown, green, orange
R10	22 R × 1 W—red, red, black

Capacitors

C1, C2	10 μF/12 WVDC electrolytic
C3	0.01 μF ceramic
C4	4.7 pF ceramic
C5	10 pF ceramic
C6	0.01 μF ceramic
C7	100 μF/16 WVDC electrolytic
C8	0.1 μF ceramic
CV1, CV2, CV3	trimmers (see text)

Additional Parts and Materials

L1 to L4	coils (see text)
MIC	electret microphone, two terminals

Printed circuit board, power supply, plastic box, heat sink to Q3, wires, solder, etc.

Project 9: High-Power Push-Pull FM Transmitter

This transmitter can put in the air a 2 W signal, ranging some miles when using an appropriate external antenna.

Features

- Power supply voltage: 6 to 12 Vdc
- Frequency range: 88 to 108 MHz
- Number of transistors: 4
- Range: several miles

A high-power push-pull output stage using two transistors gives this project a performance range of up to several miles when used with an external antenna. (See Chapter 1 for a discussion of the legal restrictions related to long-range transmissions.) The output power also depends on the power supply voltage. The reader can use supplies ranging from 6 to 12 V with this transmitter. If dry cells are used for the 6 V version, they must be D or C types, as the current drain is high.

As in Project 8, the circuit needs three adjustments to be made using three trimmer capacitors. Procedures for making these adjustments can be found in the descriptions of Projects 7 and 8.

The external modulation can come from various sources such as mixers, a high-impedance microphone, CD players, or others. A mixer can be used to plug several signals into the transmitter if it is used in an experimental radio station.

The operational frequency is determined by the coils, and the reader has some options as shown below:

50 to 80 MHz	
L1	6 turns
L2	4 + 4 turns enlaced with L1
L3	6 + 6 turns
L4	5 turns enlaced with L3
80 to 120 MHz	
L1	4 turns
L2	3 + 3 turns enlaced with L1
L3	4 + 4 turns
L4	4 turns enlaced with L3

All the coils are wound around a pencil as shown in Project 8. The wire can be the AWG 20 to 24 (enameled or plastic covered).

Figure 2.42 shows the schematic diagram of the transmitter. All the components are placed on a printed circuit board as shown in Fig. 2.43. Pay special attention to the placement of the coils.

The transistors can be replaced by equivalents. Q3 and Q4 must be mounted on small heat sinks if the circuit is powered from a 12 V supply.

The recommended trimmers capacitors can be porcelain or plastic types with capacitance ranges from 2–20 to 4–40 pF.

The small capacitors must be ceramic types. The electrolytic capacitors are rated to voltages up to 16 V or as specified in the parts list.

If an external antenna is used, it must be wired to the circuit using a coaxial cable. The trimmer potentiometer is used to adjust the modulation according to the audio signal source.

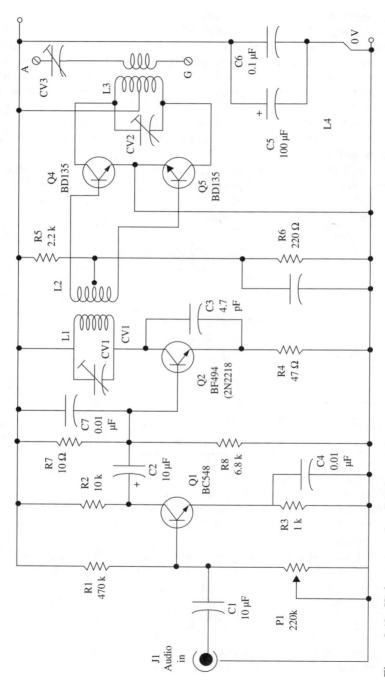

Figure 2.42 High-power push-pull FM transmitter.

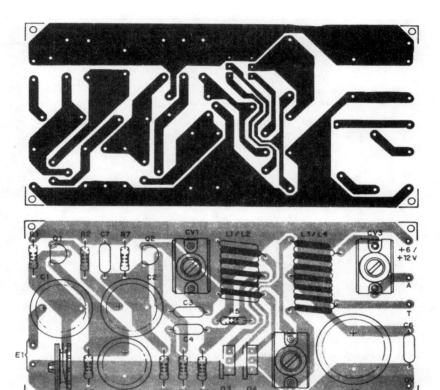

Figure 2.43   Printed circuit board for Project 9.

Parts List: Project 9

Semiconductors

Q1	BC548 or equivalent general-purpose NPN silicon transistor
Q2	BF494 or equivalent high-frequency (RF) NPN silicon transistor
Q3, Q4	BD135, BD137, or BD139 medium-power NPN silicon transistor

Resistors (1/8 W, 5%)

R1	470,000 Ω—yellow, violet, yellow
R2	10,000 Ω—brown, black, orange
R3	1,000 Ω—brown, black, red
R4	47 Ω—yellow, violet, black

Parts List: Project 9

R5	2,200 Ω—red, red, red
R6	220 Ω—red, red, brown
R7	10,000 Ω—brown, black, orange
R8	6,800 Ω—blue, gray, red
P1	220,000 Ω trimmer potentiometer

Capacitors

C1, C2	10 µF/16 WVDC electrolytic
C3	4.7 pF ceramic
C4	0.01 µF ceramic or metal film
C5	100 µF/16 WVDC electrolytic
C6	0.1 µF ceramic
C7	0.01 µF ceramic
CV1, CV2, CV3	trimmers (see text)

Additional Parts and Materials

L1, L2, L3, L4	coils (see text)

Printed circuit board, plastic box, input jack, power supply, etc.

Project 10: Pirate FM Station

This high-quality, small FM transmitter can be used as an experimental or domestic pirate broadcast station. You can produce your own radio programs and send them to receivers placed in other rooms or in your neighbors' homes.

Features

- Mixer included
- VU meter included
- Frequency range: 88 to 108 MHz
- Range: 150 ft (typical)
- Power supply voltage: 6 V (4 AA cells)

The circuit includes a high-performance mixer and VU meter and is powered from common AA cells. The signals can be tuned in by receivers placed at distances up to 150 ft. Of course, the short range is specified to avoid problems with the law (see Chapter 1 for more information).

The circuit is recommended for readers who are looking for a good project for a school science fair or a technology education project, or as an experimental radio station at school or home. Latent talent for a career as a broadcast journalist or DJ can be revealed with this simple, low-cost project. School basketball or football games can be broadcast using this station. Another use for this experimental radio station is as a wireless microphone mixer, transmitting sounds picked up by several microphones to a powerful audio system as suggested by Fig. 2.44.

We can sum up this transmitter's principal features as follows:

- Range: 150 ft
- Powered from four AA cells or supply plugged to the ac power line
- Three audio inputs with independent volume controls
- High-gain mixer stage, allowing operation with low-signal sources as microphones
- Added VU meter
- Modulation adjustment available for better performance

How It Works

The audio input circuit is formed by a mixer using a junction field effect transistor (JFET) and a bipolar common transistor. The configuration produces a large gain from the input signals, with high fidelity. Even signals from low signal sources such as low-impedance microphones and guitar pickup transducers can be used with this circuit. The original circuit was designed to use three inputs, but the reader is free to add more. Associated with each input, there is a potentiometer to control the signal level applied to the amplifier stage.

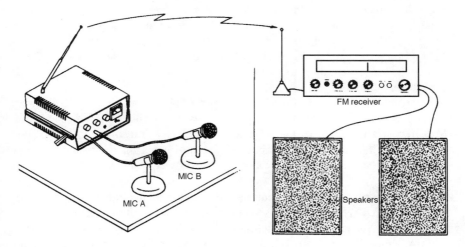

Figure 2.44   The transmitter can send signals to a remove receiver.

By adjusting the potentiometers, it is possible to fix the signal level of each input and mix them to form the transmitted signal. For instance, you can adjust the level of a microphone signal to be higher than the level of music coming from a CD player and transmit words along with the musical background.

The signals produced by this circuit are applied to the input of a small FM transmitter using a BF494 transistor. (Notice that the signal also can be applied to the input of powerful transmitters stages, but the reader must make sure that it will not cause legal problems as discussed in Chapter 1.)

The transmitter is adjusted by CV to operate at a free point of the FM band. The antenna is a telescoping type, 10 to 30 inches long, but a long piece of plastic-covered rigid wire can also be used.

If the reader lives on a farm or in another sparsely populated area, an external antenna can be used experimentally. The reader can also replace the transistor with a 2N2218 and power the circuit from 9 to 12 V supplies. Resistor R11 must be replaced by a 2 W unit. Using an external antenna, and with the modifications described above, it is possible to send the signals over distances up to a mile.

The audio output signal is also sent to a VU meter formed by transistors Q4 and Q5. The transistors amplify the signal and apply it to a microammeter.

Two components found in this stage can be altered to obtain best performance. R12 is chosen according to the VU meter sensitivity and can assume values between 4,700 and 47,000 Ω. Otherwise, R8 must be reduced to obtain the zero adjustment. Of course, this stage is optional and can be omitted if the reader prefers.

The power supply is formed by four cells (AA, D, or C). If the reader prefers to use an ac power supply, it is important to provide a good filtering circuit to avoid hum and noise.

Assembly

The schematic diagram of the pirate FM station is shown in Fig. 2.45. The components are placed on a printed circuit board as shown in Fig. 2.46. All the pieces can be installed into a plastic box as suggested by Fig. 2.47. Boxes like the one used in the prototype can be obtained from a dealer or fabricated by the reader. Notice that the potentiometers and the input jacks must be wired using shielded cable to avoid hum.

The recommended field effect transistor is the BF245, but the MPF102 can be used. If you make this substitution, observe that the terminal placement for the MPF102 isn't the same as the original. The reader can also use a 2N2218 or 2N2222, but take care when placing these equivalents on the printed circuit board. The other transistors also can be replaced by equivalents, as the circuit is not critical.

For this prototype, we use a common linear rotating potentiometer as shown in the figures but, if you prefer, it is possible to use slide potentiometers instead.

For M1 we recommend any microammeter with scale range between 200 and 500 μA. Remember that this is a polarized component, so if the needle tends to indicate negative audio levels, you must reverse the connections.

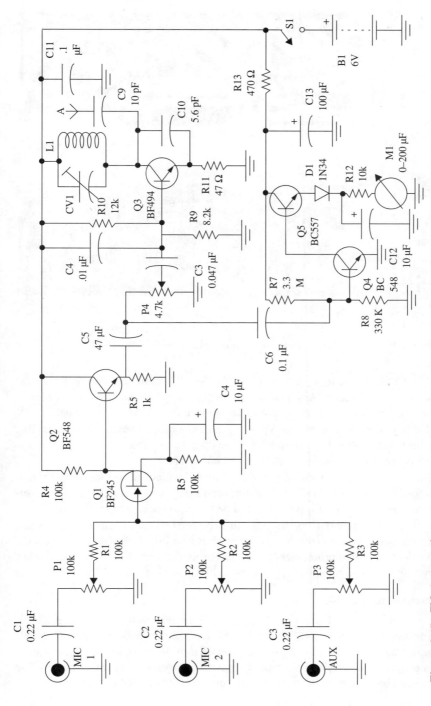

Figure 2.45   FM pirate station.

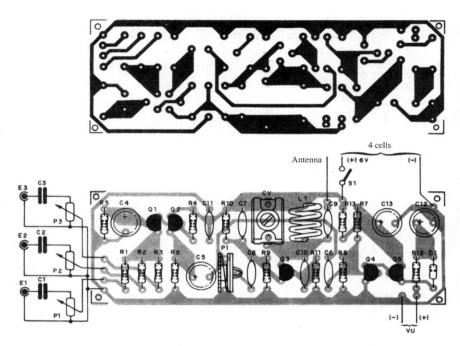

Figure 2.46   Printed circuit board pattern recommended for Project 10.

Coil L1 is formed by four turns of AWG 18 to 22 enameled wire on a coreless form 1 cm in diameter. Any plastic or porcelain trimmer capacitor can be used. This component is not critical, and values between 20 to 40 pF can be used.

It is important to observe the types of capacitors recommended in the parts list. The units used in the high-frequency stage must be ceramic. The electrolytic capacitors are rated to working voltages of 12 V or more.

For the audio input signals, we use P2 or RCA jacks according to the signal sources. The microphones are common high-impedance types. If very low-impedance microphones are used, it would be better to use a preamplifier stage.

If this transmitter is to be used as an experimental radio station, we suggest that you fabricate two extension cables as shown in Fig. 2.48. These will be used to connect remote microphones. To connect tape recorders, CD players, and other signal sources, you must use appropriate cables and plugs.

Adjustments and Use

Put the cells into the battery holder and place the transmitter close to an RF receiver tuned to a free point of the FM band. The receiver must be adjusted to half volume, and you must plug a program source into the input of the radio station. This program source can be a tape deck with a music tape or a CD player. The signal can be picked up from the speaker or earphone outputs in those devices.

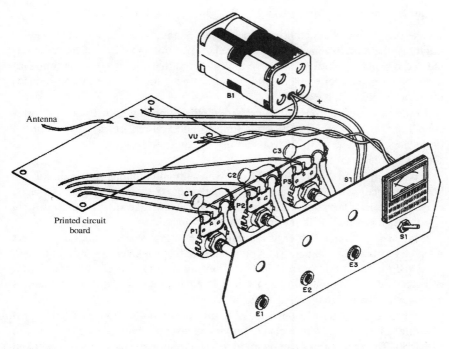

Figure 2.47   Panel detail showing placement of the input jacks and control potentiometers.

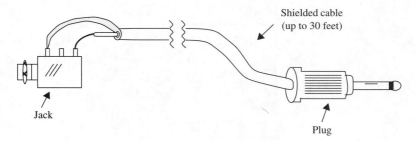

Figure 2.48   Extension cable to be used with remote sound sources.

Turn on S1 in the transmitter and adjust CV1 to tune the signal to match the receiver setting. Take care to not tune in harmonics or spurious signals.

When you are receiving the signal, adjust the corresponding input potentiometer to two-thirds of its maximum, and set P4 so that you have a strong and clear audio signal in the receiver. Then, observe whether the VU-meter is indicating the signal. Replace R8 if necessary to achieve better performance.

Radio Station Operation

To use the circuit in an experimental low-power radio station, we must keep in mind that the three potentiometers are used to determine the level of the transmitted audio signal. Therefore, we have to plug the signal sources into the transmitter's inputs and experiment to find the adjustment for each one that produces undistorted sound in the receiver.

To pass the signal from one source to another, you can adjust the two corresponding potentiometers at the same time. While you reduce the volume of the first source with one of them, you can use the other to increase the volume of a second source.

It is important to use a small FM receiver with an earphone as a monitor. Do not use a receiver with an operating loudspeaker. If the microphone is placed close to the receiver's speaker, acoustic feedback will result in the form of a strong whistle, which will interfere with your transmission.

When using the transmitter as a public address (PA) system, the loudspeakers must be located in a manner that avoids acoustic feedback, as shown in Fig. 2.49. Also keep in mind that open-field operations allow you to transmit over greater distances than if you place the transmitter in a location where metallic structures can interfere with signal propagation.

If you don't have a microphone to plug into your radio station, you can improvise one using a small loudspeaker and an audio output transformer as shown in Fig. 2.50. The transformer can be rated to primary impedances between 200 and 2,000 Ω.

If distortion appears when using CD players or tape recorders (when the signal is picked from the earphone output), it will be necessary to wire a 47 $\Omega \times$ 1/2 W resistor in parallel with the audio output cable.

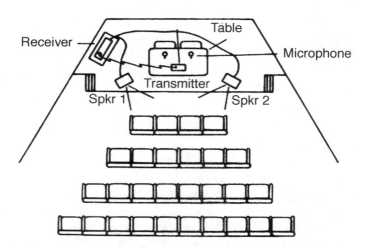

Figure 2.49   Using the transmitter as a public address system in an auditorium.

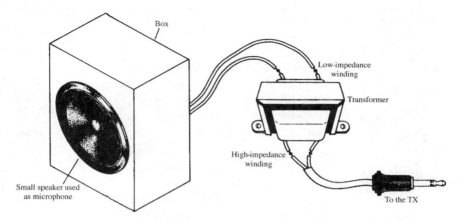

Box

Low-impedance winding

Transformer

High-impedance winding

Small speaker used as microphone

To the TX

Figure 2.50 A small loudspeaker with a transformer can be used as a microphone.

Parts List: Project 10

Semiconductors

Q1	BF245 or equivalent JFET (see text)
Q2, Q4	BC548 or equivalent general-purpose NPN silicon transistor
Q3	BF494 or equivalent small-signal RF silicon transistor
Q5	BC558 or equivalent general-purpose PNP silicon transistor
D1	1N34 or equivalent general-purpose germanium diode

Resistors (1/8 W, 5%)

R1, R2, R3	100,000 Ω—brown, black, yellow
R4, R12	10,000 Ω—brown, black, orange
R5	4,700 Ω—yellow, violet, red
R6	1,000 Ω—brown, black, red
R7	3,300,000 Ω—orange, orange, green
R8	330,000 Ω—orange, orange, yellow
R9	8,200 Ω—gray, red, red
R10	12,000 Ω—brown, red, orange
R11	47 Ω—yellow, violet, black
R13	470 Ω—yellow, violet, brown

Parts List: Project 10 (continued)

P1, P2, P3	100,000 Ω potentiometer
P4	4,700 Ω trimmer potentiometer

Capacitors

C1, C2, C3	0.22 µF ceramic or metal
C4, C12	10 µF/12 WVDC, electrolytic
C5	47 µF/12 WVDC, electrolytic
C6	0.1 µF ceramic or metal film
C7	0.01 µF ceramic
C8	0.047 µF ceramic or metal film
C9	10 pF ceramic
C10	5.6 pF ceramic
C11	0.1 µF ceramic
C13	100 µF/12 WVDC, electrolytic

Additional Parts and Materials

M	0–200 µA moving coil ammeter
L1	coil (see text)
CV	trimmer capacitor (see text)
A	antenna (see text)
S1	SPST toggle or slide switch
B1	6 V, 4 AA, C or D cells

Printed circuit board, plastic box, input jacks, shielded cable, wires, battery holder, knobs to the potentiometers, solder, etc.

Project 11: High-Power Varicap FM Transmitter

This transmitter is modulated by a variable capacitance diode (varicap) and can transmit a high-quality audio signal up to several miles.

Features

- Power supply between 6 and 12 V when using the 2N2218 transistor
- Power supply between 12 and 18 V when using the 2N3553 transistor

- High output power: 1 W with the 2N2218, 2 W with the 2N2553
- Current drain between 200 and 500 mA
- Wide frequency band: between 60 and 120 MHz
- Varicap modulation

High-power FM transmitters must not be used in cities or other places where they might interfere with other communication systems. However, in sparsely populated areas such as farms or campgrounds, a high-power FM transmitter can be used successfully under limited conditions (see Chapter 1 for more information). The transmitter described here can be used as an experimental radio station in such places. You can produce your own radio programs for clubs, camping trips, or just for listening at various locations on a farm.

Our project consists of a two-transistor, high-power transmitter with a push-pull output stage. It can be powered from 6 to 18 V supplies, depending on the version. This transmitter can deliver 1 to 2 watts of signal to an antenna, which is enough to cover distances up to several miles.

The principal characteristic of this project is the use of a variable capacitance diode (varicap) in the modulation circuit. Using this device, the circuit can produce a high-quality signal, which is important for use as an experimental broadcast station.

The circuit is designed to use an electret microphone, but the reader can use other signals sources, including the mixer stage described in Project 10. The circuit generates a high-power output, so it is important that the reader understand and observe laws limiting or prohibiting the operation of these devices (see Chapter 1).

How It Works

The core of the circuit is a high-frequency oscillator that uses two transistors in a push-pull configuration. The transistors are wired to a center-tapped coil, providing the feedback circuit.

The coil, with a trimmer capacitor and a varicap, determine the operational frequency. The trimmer must be adjusted to a central frequency, at a free point in the FM band.

The feedback signal to Q2 is picked up from the collector of Q3 via C4, and the feedback signal to Q3 is picked up from Q2 via C5. The transistors are biased by R5, R6, R7, and R8. The values of the resistors vary according to the transistor used in the circuit. For the 2N3553, resistors R6 and R7 must be reduced to 6,800 Ω or 5,600 Ω, and R6 and R7 to 4,700 or 3,900 Ω.

The signal produced by the circuit is coupled to the antenna by a second coil interlaced with L1, the oscillator coil. An electret microphone picks up the audio signal and applies it to the base of an NPN general-purpose transistor for amplification. The amplified signal is then applied to the varicap via R4.

It is useful to discuss how the varicap diode functions in this circuit, as the same principle is applied in many other applications. Any diode acts as a varicap when it is reverse biased. What happens is that, when the junction of a common diode is reverse biased as shown in Fig. 2.51, the charge carriers are separated, and a dielectric region forms between them. This means that the device can be considered to be a capacitor.

However, the separation between the charge carriers depends on the reversed applied voltage, so we can control this virtual capacitor's capacitance by changing the applied voltage. Higher voltage increases the charge separation and so decreases the capacitance.

A common silicon diode such as the 1N4002 acts as a varicap—but not a good one, because the achieved capacitance range is narrow. If you use a common diode in a circuit like the one described here, it will operate, but only when using strong audio signals. And yet, there are special diodes with large junction surfaces that can combine a wide range of capacitances with a low range of reverse voltages. Those diodes are named *varicaps* and can be found in radio receivers, TV sets, and other tuning circuits.

Figure 2.52 shows how a varicap can be used to tune a resonant circuit using a potentiometer or trimmer potentiometer. The varicap, in series with a common capacitor, replaces the variable capacitor in a common LC resonant circuit. The coil and the varicap basically determine the resonant frequency. Normally, C is chosen so as to avoid influencing the net capacitance presented by itself and the varicap. This means a value many times higher than the mean capacitance presented by the varicap.

The voltage that changes the varicap capacitance is applied via R, coming from a trimmer or common potentiometer, so the resulting capacitance and the circuit resonant frequency can be changed by the trimmer or common potentiometer. Many FM, TV, and communication receivers are tuned by such circuits.

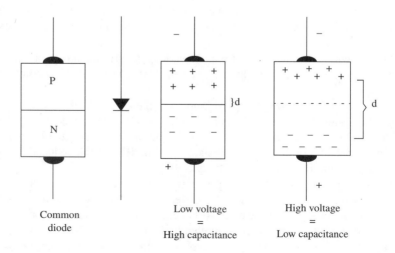

Figure 2.51   How a diode can be used as a variable capacitor.

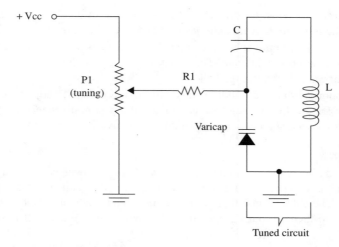

Figure 2.52   Using a varicap to tune a resonant circuit.

To generate the modulation from an audio signal, it is enough to apply this signal to the varicap. The voltage changes produced by the audio signal will change the varicap capacitance, making the frequency rise and fall within a narrow band with center frequency tuned by the trimmer capacitor as shown in Fig. 2.53.

It is important to observe that any signal applied to the varicap will modulate the high-frequency signals in frequency, including stereo multiplexed ones. Therefore, the circuit described in Project 12 intended to produce an FM multiplex stereo signal can be coupled to this transmitter with no problems.

To drive the transmitter from power supplies plugged to the ac power line, it is important to provide a good, clean circuit. Power supplies must have good filtering and stabilization to avoid noise and hum. Powering the circuit from a battery

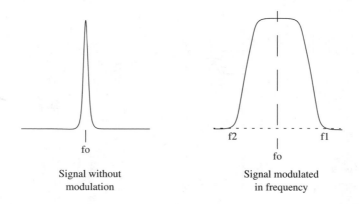

Figure 2.53   The signal band spreads when modulated by a low-frequency signal.

avoids these problems, but the battery must be able to handle the high current drain of the transmitter. Auto batteries or nicads are recommended for this task.

We must also remind the reader that the range depends not only on output power but also on the antenna. Three types of antennas that can be used to increase range are shown in Fig. 2.54.

It is important to use a standing wave ratio (SWR) meter to adjust the antenna and cable length. In this way, you can obtain the best performance by avoiding stationary waves that reduce the output power and overload the output transistors.

Assembly

Figure 2.55 shows the complete diagram of the high-power varicap transmitter. The principal components are mounted on a printed circuit board as shown in Fig. 2.56. The output transistors must be mounted on heat sinks, as their temperature tends to increase during operation. The pinout to the 2N3553 is shown in Fig. 2.57.

There are many varicaps that can be used in this project. The author recommends in particular the BB809, BB909, BB405, or BB106. The reader is free to conduct experiments to find the one that provides the best performance. One can find good varicaps in an old key-touch tuned TV set, inside the tuning unit.

The capacitors have types and voltage rates as recommended in the parts list. The resistors also have their specifications given in the parts list. If you intend to use external audio sources instead of the electret microphone, you only have to remove this component, and also R1, and apply the signal directly to C1.

To avoid distortion with high-level signal sources, place a 470,000 Ω trimmer potentiometer between the base of Q1 and the ground, and adjust the circuit for

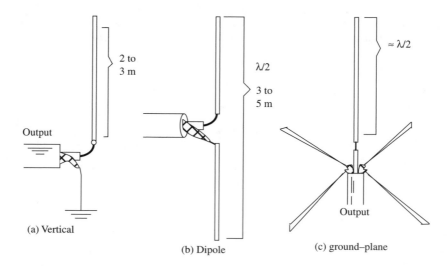

Figure 2.54   External antennas.

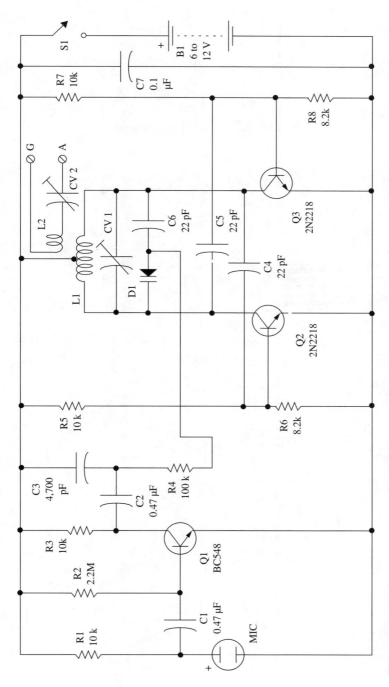

Figure 2.55   High-power varicap transmitter.

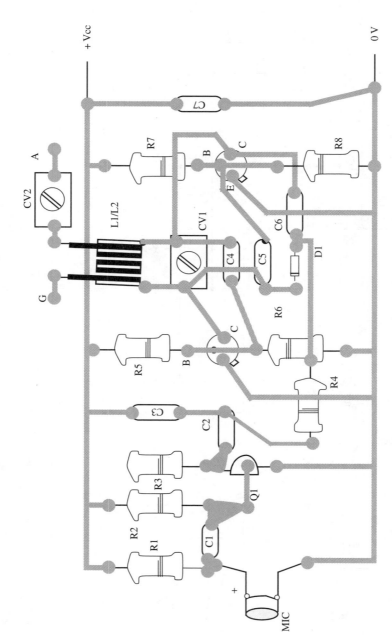

Figure 2.56 Printed circuit board for Project 11.

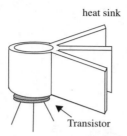

Figure 2.57   Q2 and Q3 must be mounted with heat sinks.

best performance. Any trimmer capacitor with capacitances in the range between 2–20 and 4–40 pF can be used.

L1 is formed by 3 + 3 turns of AWG 20 to 22 enameled wire wrapped on a 1 cm diameter coreless form (air core). L2 is formed by 3 or 4 turns of the same wire interwoven with L1. You can use a pencil as a reference for winding these coils. Common plastic covered wire can also be used.

A power supply to this circuit is shown in Fig. 2.58. This circuit has a 12 V output with a current up to 1 A. You can also use a 7815 to increase the output voltage to 15 V. The IC, in both cases, must be mounted on a heat sink. The circuit diagram shows the ICs that can be used and the corresponding voltage outputs.

As the power supply circuit is not critical, you can use a terminal strip as its chassis, as shown in Fig. 2.59. The transformer has a primary winding rated to 117 Vac, and a center-tapped 12 V secondary rated to 1 A, if the circuit is the 12 V version. For the 15 V version, use a 15 V CT transformer.

C1 must be rated to 25 V or more and C2 to 16 V or more. C3 must be ceramic. If the power supply is mounted as an external unit, it must be connected to the transmitter by a shielded cable.

Adjustments and Use

Near the transmitter, place an FM receiver tuned to a free point in the FM band. The distance between the units must be at least 6 ft. Use a small (8 to 20 inches long) antenna to make the adjustments. This antenna can be made of rigid wire.

Adjust CV to tune the strongest signal sent by the transmitter. Take care that you do not tune to harmonics. If the receiver output includes a strong hum, check to see that the power supply is correctly plugged in. It is recommended that you install the transmitter and power supply in metal boxes, which will shield against hum and other noise. In some cases, you can reduce hum by installing a 0.1 µF × 600 V capacitor in parallel with the ac power line.

If you find it difficult to tune the circuit because the transmitter is not operating at the desired frequency, you can alter the number of turns in L1. Conduct experiments to find the best performance. This may be necessary to compensate for component tolerances and characteristic variations introduced by a particular mounting layout.

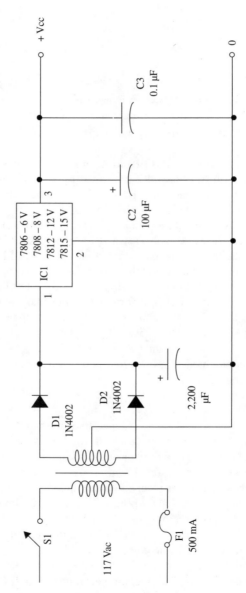

Figure 2.58 Power supply suitable for this transmitter.

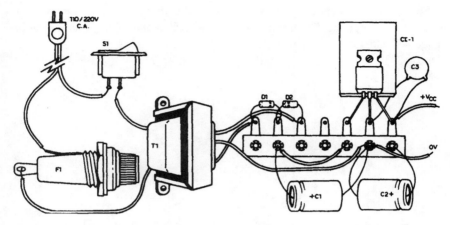

Figure 2.59   The power supply can use a terminal strip as chassis.

After determining the best operational configuration using the small antenna, you can wire the circuit to an external antenna. Make readjustments to again establish the best performance.

A simple multimeter can be used as a field strength meter as shown in Fig. 2.60. The transmitter's trimmer capacitor must be adjusted to highest voltage indication on the multimeter. The multimeter and antenna must be placed 3 or 4 ft away from the transmitter's antenna.

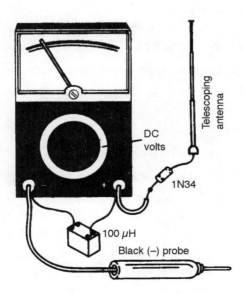

Figure 2.60   A multimeter used as a field strength meter.

Figure 2.61 shows how we can wire two external inputs to audio sources while using the microphone. The connections to the switch and input jacks should be made with shielded cable to avoid hum.

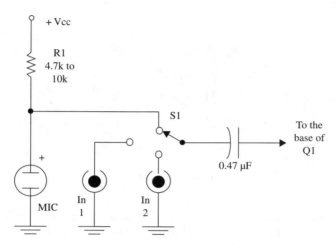

Figure 2.61   Adding external inputs to the circuit.

Parts List: Project 11

Semiconductors

Q1	BC548 or equivalent general-purpose NPN silicon transistor
Q2, Q3	2N2218 or 2N3553 RF transistor (see text)
D1	BB809 or equivalent variable capacitance diode (varicap) (see text)

Resistors (1/8 W, 5%)

R1, R2, R5, R7	10,000 Ω—brown, black, orange
R2	2,200,000 Ω—red, red, green
R4	120,000 Ω—brown, red, yellow
R6, R8	8,200 Ω—gray, red, red

Capacitors

C1, C2	0.47 µF ceramic or metal film
C3	4,700 pF ceramic

Parts List: Project 11 (continued)

C4, C5	22 pF ceramic
C6	15 pF ceramic
C7	0.1 µF ceramic
CV1, CV2	trimmers (see text)

Additional Parts and Materials

MIC	electret microphone (two terminals)
L1, L2	coils (see text)
S1	SPST toggle or slide switch

Printed circuit board, heat sinks for the transistors, shielded cable, metallic box, terminal strip to antenna/ground, telescoping or external antenna, power supply, wires, solder, etc.

Project 12: Stereo FM Transmitter

Stereo multiplexed signals can be sent with this simple transmitter. This is a real experimental stereo broadcast station that can be used for your school, club, or home.

Features

- Power supply: 6 to 12 Vdc
- Range: 150 to 600 ft
- Input sensitivity: 200 mV (peak to peak)
- Frequency range: 88 to 108 MHz
- Pilot signal: 19/38 kHz

The directional sensation we experience when listening to stereo sound is the result of channel separation. Two program sources are separated and reproduced by two different loudspeakers. This means that sounds reach our ears from different directions, causing a directional sensation.

In common audio equipment such as CD players, tape decks, and other components, the separation between the two signal sources is easily achieved, as the sounds originate from two different channels. The two signals are applied to two separate amplifiers from which we obtain reproduction through two loudspeakers or stereo earphones.

However, if we want to transmit two audio signals using one radio carrier, the problem becomes more complex. How can we modulate the high-frequency signal with two audio low-frequency signals without mixing them? And if they are mixed for transmission, how can we separate them in the receiver?

To transmit two different audio signals via one high-frequency carrier, we need a special kind of modulation circuit. This kind of modulation is named *multiplex,* and the circuit described here, called a *multiplexer,* is designed for this task.

The process used to accomplish the multiplexing of two audio signals for FM transmission follows an international standard. It is therefore important to observe that because common FM stereo receivers can recognize only signals that have been coded according to that standard, other multiplexing processes will not work. This means that there must be a correct match between the two systems if we want to achieve stereo reproduction of a transmitted stereo signal (see Fig. 2.62).

Our circuit can multiplex stereo signals coming from any stereo source, including tape decks, CD players, two microphones, etc. The reader can use the multiplex circuit with other transmitters such as described in this book, and all of them can send the multiplexed signal without encountering any problems. You only need to wire the multiplexer between any transmitter and a stereo audio source to achieve conventional stereo transmission.

The uses suggested for this transmitter are the following:

- It can be used as a wireless tape deck or CD player, sending signals from a fixed source to an FM receiver in a car or to a portable (Walkman®) FM receiver by the pool.
- It is useful in experimental radio stations for schools or clubs, sending various stereo programs to portable receivers.

How It Works

Figure 2.63 shows a block diagram of our stereo transmitter. The multiplex oscillator runs at 76 kHz, and this circuit provides the synchronism signal to the mul-

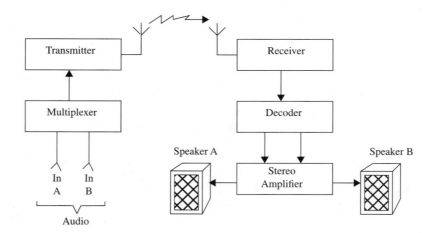

Figure 2.62 Block diagram of the stereo receiver/transmitter system.

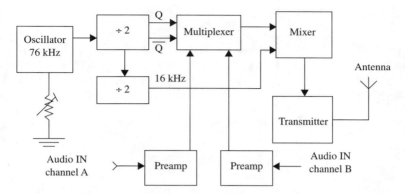

Figure 2.63   Block diagram of the stereo transmitter.

tiplexing process. The configuration is based on an RC circuit to allow easy adjustments. The RC circuit is wired to one of the four gates existing in a CMOS IC 4093. This circuit must be adjusted by P1 to run at 76 kHz.

The signal produced by this oscillator is a square wave and is applied to the input of the other gate of the 4093 IC, which is used as a buffer/inverter. From this point, the output of the circuit is applied to a 4013 CMOS IC that consists of a dual J flip-flop. This circuit divides the frequency by two and four, providing two in-phase opposition 36 kHz signals and a 19 kHz signal.

The multiplexing circuit that follows this stage is based on two of the existing analog/digital switches of a 4066 IC. You can also use the equivalent IC 4016.

The 4066 is better than the 4016 IC, as the resistance in the on state presented by the switches is lower. This means less attenuation to the audio signals that must be multiplexed, but the difference is not significant when working with strong audio signals.

The switches are electronically activated at the frequency of 38 kHz. So, in the first cycle, when one switch is closed, letting the signal pass from channel A, the other is open, blocking signals from channel B. In the next cycle, the first switch opens, and the signal from channel A is blocked, enabling the signal from channel B. This is shown in Fig. 2.64. This means that the signals are cut and applied to the high-frequency carrier but are not mixed; their parts transmitted side-by-side. This means that, using the "inverse" circuit in the receiver, it is possible to recover the original signal of each channel.

The multiplexed signals to be applied to the transmitter stage must be mixed to a 38 kHz signal coming from R2, synchronizing the receiver. This is the signal that, when recognized by the receiver, makes the stereo LED light up. The signal coming from the audio sources passes by a simple one-transistor amplification stage.

To avoid problems with high-frequency switching, which in some cases can generate noise in the receiver, it is necessary to add a lowpass filter to the circuit.

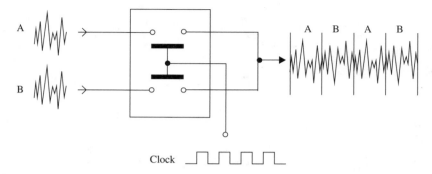

Figure 2.64   The output signal is composed of parts of channel A and B signals.

This is simply a capacitor, from 100 pF to 0.022 μF, wired between R14 and the ground.

The high-frequency stage that produces the radio signal is a simple one-transistor oscillator. We are using the BF494 as the core of this circuit, but the reader can replace it with more powerful stages.

Some components in this stage have values that depend on the power supply voltage. The values for a 12 V supply are in parentheses. The others are indicated for a 6 V power supply.

Assembly

Figure 2.65 shows the schematic diagram of the transmitter. The components are placed on a printed circuit board as shown in Fig. 2.66. The use of DIP sockets is recommended for mounting the integrated circuits.

The trimmer potentiometer is a common variety, but for more accurate adjustment, you can use a multi-turn type with proper alteration of the printed circuit board layout. The potentiometers are linear or logarithmic and must be placed in the transmitter's front panel for easy access.

The coil is formed by four turns of AWG 18 to 22 enameled wire over a coreless form 1 cm in diameter. You can use a pencil as the reference to wire this coil.

Any porcelain or plastic trimmer capacitor in the range between 2–20 to 4–40 pF can be used here. The reader must measure the distance between the component's terminals and alter the printed circuit layout if necessary. For this reason, it is best to have the trimmer in hand before etching the printed circuit board.

Any piece of rigid wire between 8 and 40 inches in length can be used as the antenna. You can also use a common telescoping antenna found in any non-operating transistor radio. The antenna can be wired to different taps in the coil, as described in previous projects, to find the best performance.

For audio input, RCA jacks are used or as dictated by the signal sources. The jacks must be wired to the printed circuit board and potentiometer using shielded

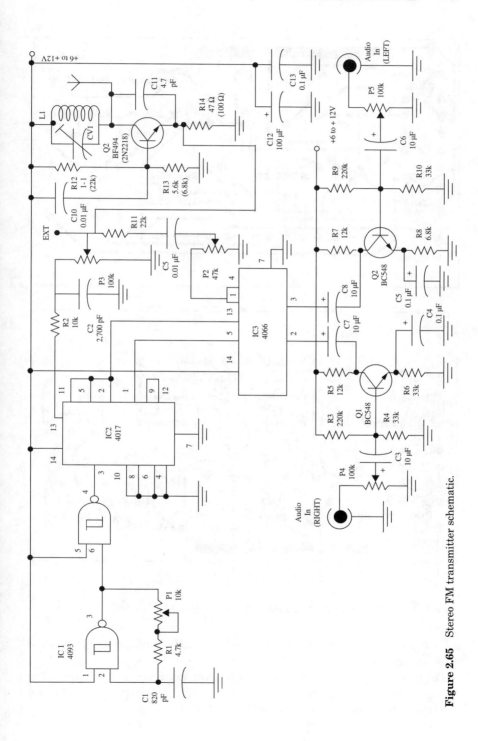

Figure 2.65 Stereo FM transmitter schematic.

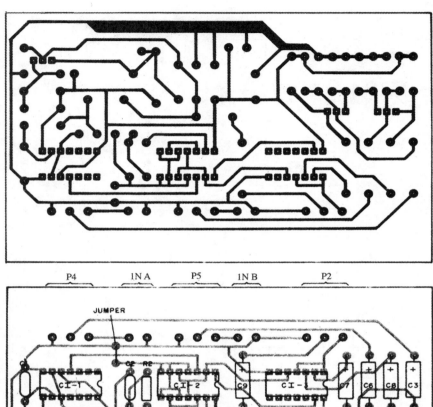

Figure 2.66   Printed circuit board suggested for Project 12.

cable to avoid hum. You can also use a stereo jack if the plug used to apply the audio signals is compatible.

The power supply can be AA cells or a battery, depending on the version (6 or 12 V). If you have to use an ac-powered supply line, it is important to obtain a power supply that offers good filtering and regulation.

If you intend to use the transmitter for long time periods, it is important to increase the battery capacity. We recommend the use of C or D cell in this case. A suitable plastic box for installing the unit is suggested by Fig. 2.67.

Adjustments and Use

It is important to have available an FM stereo receiver with a stereo LED indicator. First, we tune the receiver to a free point in the FM band. The receiver must be placed 3 or 4 ft away from the transmitter antenna.

Second, we turn on the transmitter and adjust the trimmer capacitor to tune the signal. Take care to tune in the fundamental and not harmonic signals. Next, we need to adjust P3 and P2 to the medium point of their range.

After this, we adjust trimmer potentiometer P1 until the stereo LED in the receiver glows. At this point, the multiplex circuit is operating in 38 kHz as needed by the receiver's decoder stages.

Next, slowly reduce the pilot signal amplitude by adjusting P2 while you fine tune the trimmer setting. The ideal point in this adjustment is when you obtain a narrow band tuned by the receiver in which the stereo LED is on. At this time, we can plug a stereo source, such as a tape deck, signal generator, or two microphones, into the inputs. By opening potentiometers P4 and P5 and also adjusting P2, we obtain the best stereo reproduction in the receiver.

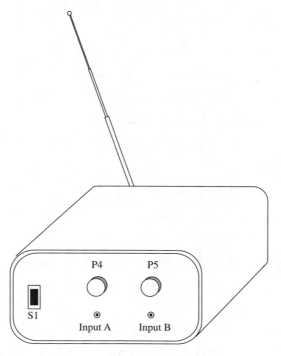

Figure 2.67   The circuit can be housed in a plastic box.

If high-frequency noise remains in the transmission, we must wire a capacitor in the range of 0.01 to 0.1 μF between the emitter of Q4 and ground. After all these adjustments are complete, we can use the transmitter. While in use, none of the adjustments can be changed, except P4 and P5, which can be adjusted according to the signal source power.

The multiplexer stereo encoder can be used with high-power transmitters. The signal must be picked up from the *external modulation point* shown in the diagram. In this case, the transmitter stage must be deactivated, and R14 must be removed from the circuit, as is transistor Q3.

Parts List: Project 12

Semiconductors

IC-1	4093 CMOS integrated circuit, four NAND Schmitt gates
IC-2	4013 CMOS integrated circuit, dual J flip-flop
C1-3	4066 or 4016 CMOS integrated circuit, 4 analog/digital CMOS switches
Q1, Q2	BC548 or equivalent general-purpose NPN silicon transistor
Q3	BF494 high-frequency NPN transistor

Resistors (1/8 W, 5%)

R1	4,700 Ω—yellow, violet, red
R2	10,000 Ω—brown, black, orange
R3, R9	220,000 Ω—red, red, yellow
R4, R10	33,000 Ω—orange, orange, orange
R5, R7	12,000 Ω—brown, red, orange
R6, R8	6,800 Ω—blue, gray, red
R11	22,000 Ω—red, red, orange
R12	8,200 Ω—gray, red, red
R13	5,600 Ω—green, blue, red or 8,200 Ω—gray, red, red
R14	68 Ω—blue, gray, black or 100 Ω—brown, black, brown
P1	10,000 Ω—trimmer potentiometer
P2	47,000 Ω—potentiometer
P3	100,000 Ω—trimmer potentiometer
P4, P5	100,000 Ω—potentiometers

Parts List: Project 12 (continued)

Capacitors

C1	680 to 820 pF ceramic
C2	2,700 pF ceramic
C3, C6, C7, C8, C9	10 µF/12 WVDC, electrolytic
C4, C5	0.1 µF ceramic
C10	0.01 µF ceramic
C11	4.7 or 5.6 pF ceramic
C12	470 µF/12 WVDC, electrolytic
C13	0.1 µF ceramic
CV	trimmer capacitor (see text)

Additional Parts and Materials

L1	coil (see text)

Printed circuit board, input jacks sockets, knobs to the potentiometers, shielded cables, wires, plastic box, battery holder or power supply, telescoping antenna, solder, etc.

Project 13: FM Super Transmitter

This high-power transmitter uses a push-pull output stage with a high-frequency transistor to drive output powers up to 5 W. The signal transmitted by this transmitter can be received at distances up to many miles.

Features

- 12 to 15 V power supply voltages
- Current drain: up to 1.5 A
- Frequency range: 88 to 108 MHz
- Output power: 2 to 5 W
- Modulation: using varicap
- Audio input sensitivity: 1 Vpp
- Output stage: class C push-pull
- Range: up to several miles

If you are looking for a high-power transmitter for experimental purposes, or even to create a broadcasting station for your home, a farm, a club, or an isolated school, this project is recommended.

Once again we must remind the reader about existing laws governing the operation of this kind of equipment. Be sure that the operation of this transmitter will not interfere with other systems or infringe on the FCC rules (or the rules that exist in your country—see Chapter 1 for more details).

This transmitter uses four transistors and operates in the FM range between 88 and 108 MHz. It is also possible to change the coils for operation in the low VHF band between 50 and 88 MHz, but the reader must have a special receiver to tune the signals and still must be careful to observe the rules related to this kind of transmission. The power output can rise to values up to 5 W due to the powerful push-pull stage, which uses special RF transistors.

It is also important to draw the reader's attention to the fact that this circuit is more critical than the others previously described in this book. Therefore, only a reader who is highly experienced with high-frequency circuits involving printed circuit board layout and tuned circuit adjustments should attempt to build this transmitter.

How It Works

When trying to diagnose problems in a nonfunctioning transmitter, is very important to know how it works. This is one of the reasons why the "How It Works" section is included in all of these projects, and readers are advised to pay special attention.

Figure 2.68 shows the block diagram representing the transmitter. The first block represents a high-frequency oscillator where L1 and CV determine the operating frequency.

This circuit is modulated by a varicap. As described in Project 11, varicaps are variable-capacitance diodes that can change the frequency of a resonant circuit by means of an audio signal (see that project for more details). The varicap is used to modulate the high-frequency signal from an audio signal picked up from an external source.

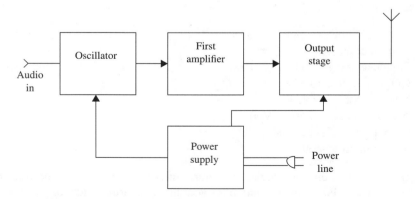

Figure 2.68 Block diagram of the FM super transmitter.

To get the best modulation, P1 must be adjusted according to the audio source output power. The modulated high-frequency signal produced by this circuit, between 88 and 108 MHz, is applied to the second block, a driving stage, or first amplifier stage using a medium-power RF transistor. The transistor is wired to the common emitter configuration with a resonant circuit as the load, plugged into the collector.

To maximize performance, it is very important to adjust the load resonant circuit to the correct frequency. This is done by adjusting the trimmer capacitor CV2. The amplified signals are then applied to the input of a two-transistor push-pull stage. This is the third block shown in our block diagram.

L4, which picks up the signals, has a CT (center tap) placed so that there are two signals in phase opposition to be applied to the output transistors' bases. This means that, in Class C operation, during each half-cycle of the input signal, one transistor conducts while the other remains off.

A high-frequency stage using this configuration can generate high output powers from small input signals. The principal advantage is that when one transistor is amplifying a half-cycle, the other does not drain any current, as it has been cut off.

The transistor chosen for this application is the 2N3866, which can drive several watts in a circuit such as this. The output signal is applied to a resonant circuit formed by L5 and CV3. Again, this resonant circuit can be carefully adjusted for better performance. CV4 is the last adjustment, matching the circuit impedance with the antenna. This is necessary to get the better signal transference into space.

The circuit is powered from a 12 V power supply. We recommend the use of a car battery to avoid problems with noise or hum. If you prefer to power the circuit from the ac line, the power supply must include a good filtering circuit. Figure 2.69 shows a power supply that may be used with this transmitter.

Assembly

The complete diagram of the transmitter is shown in Fig. 2.70. The components are placed on a printed circuit board as shown in Fig. 2.71.

The printed circuit board layout is a very important item to be considered in this project. The connections between components must be short to minimize capacitance losses that can affect the performance by introducing instabilities and hum.

Q1 and Q2 can be the original 2N2218 or equivalents such as the BD135. (Notice that the BD135 is an audio transistor, but it has a high transition frequency that allows high-frequency oscillation when used in circuits such as this.) The 2N3866 transistors must be mounted on heat sinks.

The coils are mounted on plastic forms such as those used in intermediate frequency (IF) and oscillator transformers. The transformers must be unmounted, and the coils are wound on plastic forms as shown in Fig. 2.72. These transformers can be found in many old, nonfunctioning transistor radios and TV sets.

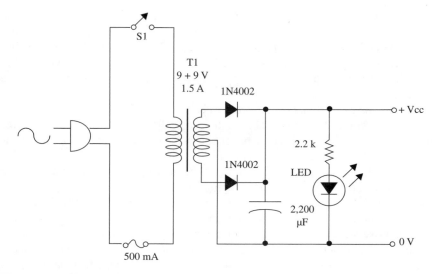

Figure 2.69   The FM super transmitter can be powered from this power supply.

Extract the transformers from the original printed circuit boards and remove the metal cover carefully. Then, remove the wire used for the original winding and wind a new coil according to the specifications given in the parts list or text.

Any plastic or porcelain trimmer with capacitance ranges between 2–20 and 4–40 pF can be used. The recommended varicap diode is the BB809 or BB909, but there are many equivalents that can be investigated. The reader can find good varicaps in the tuning circuits of antique touch-tuned TV sets or FM radios.

The electrolytic capacitors must be rated to 12 WVDC or higher. The ceramic capacitors are critical, and their values must be observed. Don't use other types where ceramic capacitors are recommended.

For the power supply, the transformer has a primary winding rated to the ac power line (117 Vac) and a secondary winding rated to 9V + CT and current of 1.5 A or more. The diodes are 1N4002 or equivalents such as the 1N4004 or 1N4007. The electrolytic capacitor for the power supply is rated to voltages of 16 WVDC or more.

It is important to install the power supply and the transmitter in separate metal boxes. Each metal box acts a shield, and they must be joined by a common ground connection.

For the audio input, use an RCA jack. The antenna must be connected to the circuit by a 75 Ω cable and connector.

Adjustments and Use

Assemble a Hertz loop, as shown in Fig. 2.73, using a small 6 V × 50 mA incandescent lamp and common 22 wire (enameled or plastic covered), and install it as shown. Place any FM receiver near the transmitter, and tune it to a free point in

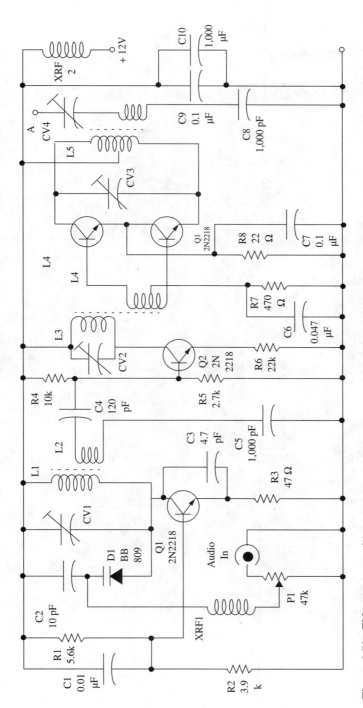

Figure 2.70 FM super transmitter.

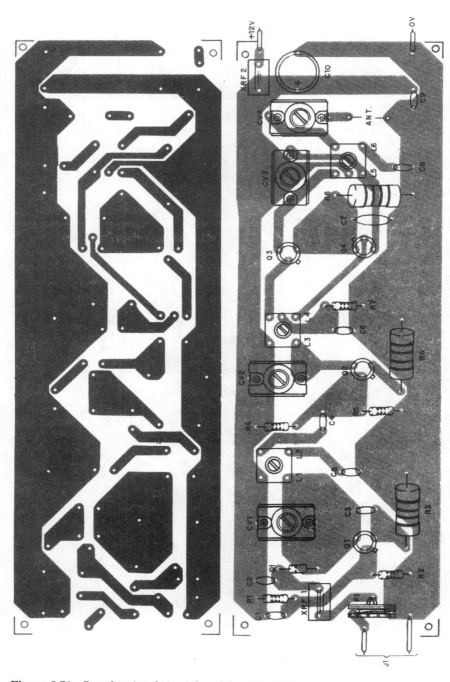

Figure 2.71 Sample printed circuit board for Project 13.

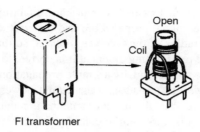

Figure 2.72   Using the form of an IF transformer to assemble the coils.

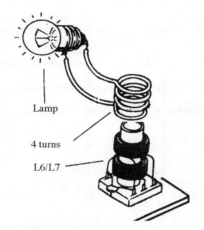

Figure 2.73   Using a Hertz loop for adjustments.

the FM band. Power the transmitter up and adjust CV1 to tune the signal in the receiver. A program source, such as a tape recorder or CD player, can be plugged into the transmitter.

The next step is adjust CV2 and CV3 to the highest output power. The lamp in the Hertz loop will glow with the highest brightness when the ideal point is found. Readjust CV1 to set the tuned signal to the exact desired frequency.

Use only a plastic or wooden tool to adjust the trimmers; metal tools can alter the capacitance by their presence, making these adjustments difficult. After adjustment, the circuit can be used with an appropriate sound source.

When using an external antenna, also adjust CV4 for best performance. For this task, you can use a field strength meter. (See the previous project to learn how a multimeter can be used as a field strength meter.)

Power and Range

As discussed occasionally in this book, many people incorrectly believe that higher transmitter power necessarily results in greater range. In fact, the range of a transmitter depends on factors that include the suitability of the antenna, the presence of obstacles to signal propagation, and many others. However, talking specifically about FM radios, if the receiver is placed so that there are no obstacles between it and the transmitting antenna, it is correct to say that more output power produces greater range. And, by a simple propagation rule, we can say that to double the range, we must quadruple the output power. This is because the amount of energy that reaches the receiver antenna is reduced by a factor of the square of the distance. It is important to consider those factors when designing a powerful transmitter, because the distance gained may not be worth the money it takes to build it.

Parts List: Project 13

Semiconductors

Q1, Q2	2N2218 or equivalent silicon NPN RF transistors (see text)
Q3, Q4	2N3866 NPN RF silicon power transistors
D1	BB809 or BB909 variable capacitance diode (varicap) or equivalent (see text)

Resistors (1/8 W, 5%)

R1	6,800 Ω—blue, gray, red
R2	4,700 Ω—yellow, violet, red
R3	47 Ω × 1 Ω—yellow, violet, black
R4	10,000 Ω—brown, black, orange
R5	3,900 Ω—orange, white, red
R6	22 Ω × 1 Ω—red, red, black
R7	470 Ω—yellow, violet, brown
R8	22 Ω × 1 Ω—red, red, black
P1	47,000 Ω trimmer potentiometer

Capacitors

C1	0.01 μF ceramic
C2	10 pF ceramic
C3	4.7 or 5.6 pF ceramic
C4	120 pF ceramic
C5, C8	1,000 pF ceramic

Parts List: Project 13 (continued)

C6	0.047 µF ceramic
C7, C9	0.1 µF ceramic
C10	1,000 µF × 16 WVDC, electrolytic
CV1, CV2, CV3	trimmers (see text)

Additional Parts and Materials

XRF1, XRF2	100 µH to 220 µH RF chokes
L1, L3	5 turns (*)
L2	4 turns*
L4	4 turns with a CT
L5	6 urns with a CT
L6	6 turns
J1	RCA jack or other according to the audio signal source
J2	coaxial jack to coaxial cable (antenna)

Printed circuit board, metal box, power supply, solder, wires, etc.

* All coils are made with AWG 22 wire in a 0.5 cm of diameter form with ferrite core. L1/L2 and L3/L4 are wound in the same form.

Project 14: FM Transmitter Using a Vacuum Tube

This transmitter uses a 6C4 tube and can transmit a powerful 4 W signal. The signal can be tuned in the frequency range between 88 and 108 MHz at distances up to some miles.

Features

- Power supply: 117 Vac
- Plate voltage to the tube: 150 V (typical)
- Output power: 2 W (typical)
- Frequency range: 88 to 108 MHz
- Range: more than 5 miles
- Modulation: FM

High-power FM transmitters are special projects that are attractive to many readers. Of course, the readers who intend to build one of them must observe all existing laws and regulations related to radio transmissions.

Although any high-power FM transmitters use transistors, the best and cheapest way to get a high power output in the FM band is via the use of a tube. Using a 6C4 tube, you can build a 4 W transmitter that operates in the range between 88 and 108 MHz. This means that the signal can carry up to several miles when the transmitter is used with an appropriate external antenna.

An important suggestion is to use the FM multiplexer stereo encoder described earlier in this chapter. This will allow the transmitter to perform as an experimental FM stereo radio station. How to modulate the transmitter signal will be explained.

The transmitter is powered directly from the ac power line. As the circuit is very sensitive to hum and noise, it is very important keep all the connections short, use shielded cables for audio and other signals, and use a metal box to contain the equipment.

Another important point to consider is that the circuit includes no isolation from the ac power line. This means that shock hazards must be considered when the circuit is in operation. All appropriate precautions must be observed to avoid such shock hazards when building and operating the transmitter.

The version described here is basic, and many improvements can be added by the reader to increase security and performance. For example, you can

- Add an isolation transformer to power the circuit, increasing physical security.
- Include a varicap modulation stage to increase the audio fidelity.
- Add filtering stages to the power supply to reduce noise and hum.
- Include a power supply that uses a high-voltage transformer to increase the tube plate voltage. This will increase the power output to more than 5 W.

How It Works

The circuit is based on a 6C4 triode tube. This tube can be found in many old radio and TV sets. Those apparatus use the 6C4 tube for several functions, i.e., as an audio preamplifier, detector, mixer, oscillator, etc. Although the 6C4 is designed for audio circuits, it can oscillate at frequencies as high as 100 MHz.

The principal characteristics of the 6C4 tube are as follows.

Filament supply voltage	6.3 V
Filament current	150 mA
Plate voltage	300 V (max)
Grid voltage	50 V (max)
Plate current	25 mA (max)
Plate dissipation power	8 W (max)
Output power with a 300 V supply	5.5 W (max)

In this project, the 6C4 tube is wired as a high-frequency Hartley oscillator in which the frequency is determined by CV and L1. Capacitor C6 provides the necessary feedback to keep the oscillations on, and R3 biases the grid. Resistor R2 biases the cathode, keeping its voltage above the grid voltage—an important condition for correct operation.

One option is to channel the audio input signals to the cathode. Applying the signal to this point, it is possible to alter the circuit frequency, thereby modulating the high-frequency signal. Another input option, recommended for use when the transmitter is modulated from weak signals sources, is to the grid.

The power supply is formed by a full-wave rectifier using four diodes and a filter using three high-value capacitors. The "pi" filter uses C3, C4, C5, and the resistor R1. This filter is very important to reduce the hum that can be superimposed on the transmitted signal as modulation. C5 must be a ceramic capacitor, as only this kind of component can be used to ground any high-frequency signal present at this point of the circuit.

To heat the filament, we use a small transformer with a secondary winding rated to 6 V × 250 mA or more.

Assembly

Figure 2.74 shows the schematic diagram of the transmitter. Although we are using a tube, it is possible to use a printed circuit board for the assembly, as shown in Fig. 2.75.

The tube's pins can be soldered directly to the printed circuit board in the configuration shown. The printed circuit board is fixed to a metal chassis. Observe that the tube's pins are numbered starting from the large space and running counterclockwise as Fig. 2.76 shows. The connections from R1 and C1 to the oscillator coil must be as short as possible to avoid hum.

Filtering capacitors C3 and C4 are not critical, and units with capacitances between 16 and 50 µF can be used. The voltage rating of these capacitors must be 200 V or more. If you intend to power the circuit from a 220 V power line or use a high-voltage transformer, the capacitors' voltage rating must be increased.

In many cases, it is possible to find dual capacitors for this task. These capacitors are assembled in one case that can be fixed in a chassis as shown in Fig. 2.77. If this type of capacitor is used, mount it on the chassis with wires connecting it to the printed circuit board. Note that the negative terminal is the case. The figure also shows how resistor R1 is wired. The printed circuit board can be altered to receive this kind of component.

Another kind of electrolytic capacitor that can be used for this project is the one shown in Fig. 2.78. This type uses a nut to fix it to a chassis (over-chassis mounting). As the negative terminal is the case, when the nut is fixed in contact with the chassis, the negative connection is automatically provided.

R1 must be a 5 W resistor. The coil is formed by five turns of AWG 18 to 24 enameled wire on a coreless form with a diameter of 1 cm.

Any porcelain or plastic trimmer with capacitances ranging between 2–20 and 4–40 pF can be used. It is recommended that you use types that are designed for high-voltage signals, such as components that use mica as the dielectric.

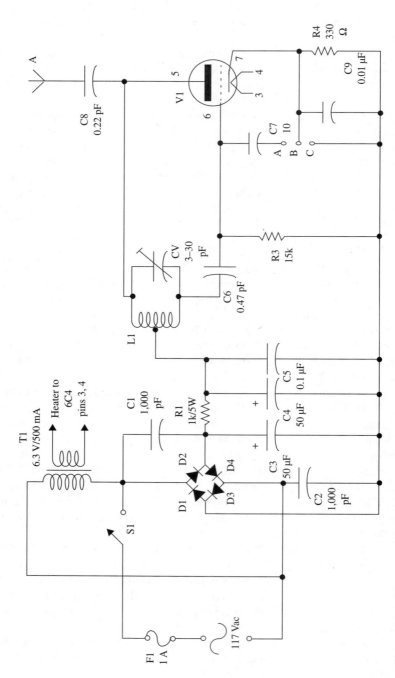

Figure 2.74 FM transmitter based on a vacuum tube.

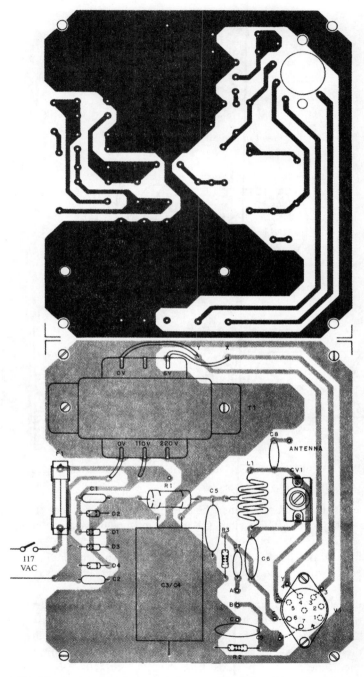

Figure 2.75   Transmitter can be mounted on a printed circuit board as shown above.

Figure 2.76 Pin placement and identification for a tube.

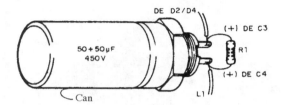

Figure 2.77 Double electrolytic capacitor.

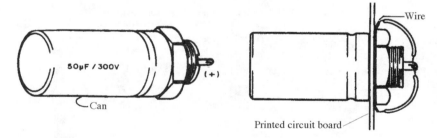

Figure 2.78 Mounting an electrolytic capacitor on a printed circuit board.

Capacitors C6, C7, and C8 must be ceramic high-voltage types. They must be rated to voltages of 400 V or more. C1 and C2 must be polyester or another metal film type rated to 400 V or more.

The transformer, used to heat the filament, has a primary winding rated to the ac power line voltage and a secondary rated to 6.3 V with currents of 250 mA or more.

If the reader uses a transformer with a low-voltage, high-current winding (6.3 V × 500 mA or more), it is possible to power a small incandescent lamp as well. This lamp can be used to indicate when the transmitter is "on the air."

The diodes are 1N4004 or 1N4007 if the circuit is powered from the 117 V power line. If the circuit is powered from the 220 V or 240 V power line or from a high voltage transformer, the diodes must be the 1N4007 or equivalents.

The printed circuit board should be fixed on a metal chassis and installed in a box to avoid accidental touch, as all metallic parts are "alive" and can be dangerous shock hazards.

Adjustments and Use

Place the transmitter near an FM receiver tuned to a free point of the FM band (between 88 and 108 MHz) and turn on the power. Use a piece of wire 5 to 10 inches long as an antenna when making these adjustments.

Wait two or three minutes for the tube to warm up. This is normal, as the cathode does not begin to emit electronics until it is heated to a relatively high temperature. In the same manner as a common incandescent lamp, the tube is heated by the electrical current. The reader should consider this high temperature to be normal and avoid touching it when it is hot.

First adjust trimmer capacitor CV1 to tune the transmitter's strongest signal to the receiver frequency. You must take care not to tune harmonics or spurious signals which, because of the transmitters high power, will exist at several frequencies in the FM band.

The modulation comes from a small audio amplifier (50 mW to 1 W) wired as shown in Fig. 2.79. The trimmer potentiometer is adjusted to find the best modulation point in which the received signal is free from distortions.

Figure 2.80 shows a circuit that can be used to achieve the modulation using a varicap. Any varicap can be used with this circuit. The ideal audio signal amplitude for best modulation depends on the varicap's characteristics.

After making all adjustments, the transmitter is ready to be used. But, as always, take care not to violate federal regulations with this high-power unit.

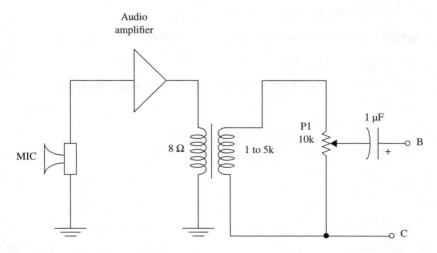

Figure 2.79 Modulation circuit.

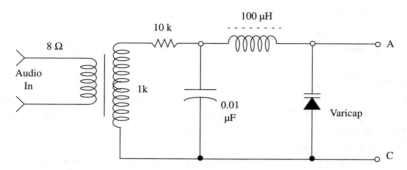

Figure 2.80   Modulation circuit with varicap.

Parts List: Project 14

Tube

6C4   miniature triode

Semiconductors (1/8 W, 5%)

D1–D4   1N4004 or 1N4007 (117 Vac power line) or 1N4007 (220 or 240 Vac power line)

Resistors (1/8 W, 5%)

R1   1,000 Ω × 5 W, wire wound

R2   330 Ω—orange, orange, brown

R3   15,000 Ω—black, green, orange

Capacitors (400 WVDC or more except electrolytics)

C1, C2   1,000 µF, metal film

C3, C4   8 to 50 µF × 200 V (117 Vac) or 400 V (220 and 240 Vac) electrolytics (see text)

C5   0.1 µF ceramic*

C6   47 pF ceramic*

C7   10 pF ceramic*

C8   22 pF ceramic*

C9   0.01 µF ceramic*

CV   trimmer (see text)

Parts List: Project 14 (continued)

Additional Parts and Materials

F1   1 A fuse and holder

S1   SPST toggle or slide switch

T1   Transformer, 117 Vac primary winding, 6 V × 250 mA secondary winding

L1   coil (see text)

Printed circuit board, metallic chassis and box, input jack, antenna jack (optional), power cord, etc.

---

*High-voltage types (400 V or more)

Project 15: Voice-Effects Transmitter

Scrambled messages can be sent with this simple transmitter. You can use this transmitter for secret conversations with your friends and neighbors.

Features

- Power supply: 6 Vdc
- Range: 300 ft
- Operation range: 88 to 108 MHz

This is a simple project that can be used for several purposes. Using a microphone as an audio signal source, it is possible to send coded messages to a remote receiver. Anyone who tunes the signal will find it difficult to understand the message. Another interesting use is to plug a guitar pickup transducer into the circuit to transmit the music, with special effects, to a remote receiver connected to a powerful audio amplifier. For readers who have already built experimental radio stations, this circuit can be useful for adding special effects to voice and music.

The circuit is powered from a 6 V supply (4 AA cells) and uses a low-power transmitter. However, the reader can replace the high-frequency stage with a high-power version, many of which are described elsewhere in this book.

How It Works

The secret of how the voice effect stage operates involves the use of a full-wave bridge that uses four diodes. This stage doubles the audio signal frequency and changes its wave shape.

The signal into this stage is picked up by a small transformer that also performs the function of isolating the circuit from the audio source. Figure 2.81 shows what happens when the audio signal is applied. As we can see, the signal

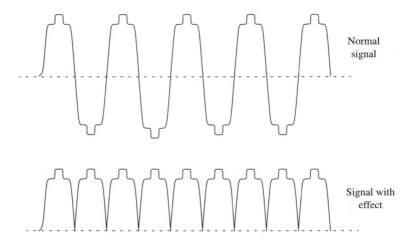

Normal
signal

Signal with
effect

Figure 2.81   The circuit changes the signal wave shape.

changes the wave shape and frequency, which means an alteration in the timbre if a voice or musical tone is used.

To drive the diode bridge, we need some power and a transformer to apply enough voltage to the diodes to break the voltage potential barrier found in the junctions. So, to drive the bridge, an audio power stage was added. It is a four-transistor amplifier with an output of about 1 W. Of course, the reader can replace this stage with any audio output circuit, including modern versions using ICs such as the LM386, TDA7052, or TBA820.

The transformer must have one low-impedance winding to be driven by the amplifier output, and a high-impedance winding to drive the diodes. In the proto-type, we used a power transformer with a 117 Vac primary winding and a 6 V × 250 mA secondary winding. The high-impedance winding is the 117 Vac pri-mary, and the low-impedance winding is the 6 V secondary.

The transmitter stage uses a BF494 powered from a 6 V supply to reach dis-tances up to 300 ft. The reader is free to use the transmitters described in other projects.

The trimmer potentiometer is used to get the best performance and least distor-tion with a particular audio source. The "X" point indicated in the diagram is the audio output to the scrambler circuit. This point can be connected to the input of any audio amplifier or tape recorder. The antenna is a piece of rigid wire 6 to 20 inches in length.

Assembly

Figure 2.82 shows the schematic diagram of the transmitter. The printed circuit board used to mount the components is shown in Fig. 2.83.

Component specifications are given in the parts list. Notice that the capacitors used in the high-frequency stage must be ceramic plate or disc types.

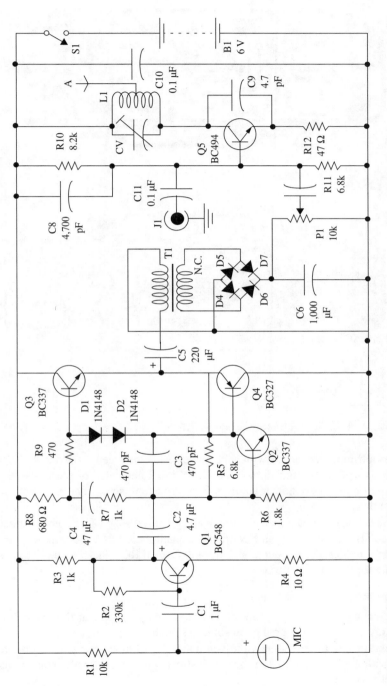

Figure 2.82   Voice effect transmitter.

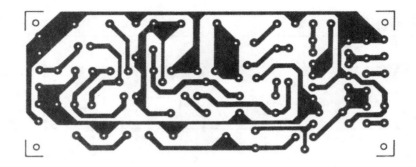

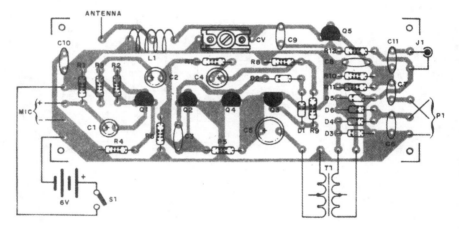

Figure 2.83   Printed circuit board for Project 15.

CV is a trimmer capacitor with capacitance range between 2–20 and 4–40 pF. The coil is formed by four turns of AWG 18 to 22 enameled or plastic-covered rigid wire on a 1 cm diameter coreless form. You can use a pencil as a form for winding this coil.

It is necessary to place the cells in a battery holder. Take care when wiring all of the polarized parts such as electrolytic capacitors, diodes, the transistor, and the battery holder. For the audio signal input, use an RCA jack or other type appropriate for the audio source.

Adjustments and Use

Tune a receiver to a free point in the FM band and place it near the transmitter. Adjust CV until you pick up the transmitted signal.

Speak into the microphone while you adjust P1 to tune your voice in the receiver with a high timbre. After these adjustments have been made, you can use the circuit as desired.

Parts List: Project 15

Semiconductors

Q1	BC548 or equivalent general-purpose NPN silicon transistor
Q2, Q3	BC337 or equivalent general-purpose NPN silicon transistors
Q4	BC327 or equivalent general-purpose PNP silicon transistor
Q5	BF494 or BF495 high-frequency NPN silicon transistor
D1–D5	1N4148 or 1N914 general-purpose silicon diodes

Resistors (1/8 W, 5%)

R1	10,000 Ω—brown, black, orange
R2	330,000 Ω—orange, orange, yellow
R3, R7	1,000 Ω—brown, black, red
R4	10 Ω—brown, black, black
R5, R11	6,800 Ω—blue, gray, red
R6	1,800 Ω—brown, gray, red
R8	680 Ω—blue, gray, brown
R9	470 Ω—yellow, violet, brown
R10	8,200 Ω—gray, red, red
R12	47 Ω—yellow, violet, black
P1	10,000 Ω trimmer potentiometer

Capacitors

C1	1 µF/12 WVDC electrolytic
C2	4,7 µF/12 WVDC electrolytic
C3	480 µF ceramic
C4	47 µF/12 WVDC electrolytic
C5	220 µF/12 WVDC electrolytic
C6	1,000 pF ceramic
C7, C10, C11	0.1 µF ceramic
C8	2,200 µF ceramic
C9	4.7 pF ceramic
CV	trimmer (see text)

Parts List: Project 15 (continued)

Additional Parts and Materials

S1	SPST toggle or slide switch
MIC	electret microphone, two terminals
B1	6 V, 4 AA cells
J1	RCA jack (see text)
L1	coil (see text)
T1	117 Vac primary winding, 6 V × 250 mA secondary (see text)

Printed circuit board, battery holder, plastic box, antenna, wires, etc.

3

Special-Purpose Transmitters

Experimental transmitters such as the ones described in this book have many uses in addition to sending voice and music to a remote receiver. The reader can find many other interesting uses. Small transmitters for FM and other frequency ranges can be used to send data picked up by sensors or transducers used in remote alarms, remote sensing applications, scientific experiments, and many other applications. This chapter is devoted to transmitters that are designed to send information and signals that have been collected by special transducers to an appropriate receiver. The range of uses is limited only by the reader's imagination.

Project 16: Noise Transmitter

This noise transmitter can be used to "cancel" or "jam" undesirable radio transmissions made by a spy transmitter hidden in a room or other location. It can be used as an efficient anti-spy device to neutralize spy transmitters such as the ones described in this book and others operating in the VHF and FM band.

Features

- Powered from the ac line
- High power, with the ability to jam signal sources over a 300 ft radius
- Wide frequency band operation, canceling signals in the range between 60 and 120 MHz
- Built with common parts
- Needs no adjustment

The best away to cancel the signal of a spy transmitter hidden in a room is to use a strong high-frequency signal to cover the signal produced by the undesirable transmitter (i.e., the "bug"). The circuit described here has just this function: it is a high-power noise transmitter that covers the FM and VHF bands and produces a strong hum in any FM receiver located within a distance of approximately 150 ft. If the spy's receiver is placed within range, it will not be able to tune in the signal of the bug, which will be overpowered by the noise transmitter.

This circuit is designed to cover transmissions in the FM and VHF band, but the reader can make changes for operation in other ranges. An interesting modification recommended for this circuit is replacing the high-frequency circuit with a 800 to 900 MHz oscillator. This is the cellular telephone frequency range. If you turn on the circuit in your building, no one will be able to use a cell phone to call out or receive messages. This allows you to use the circuit in a conference room or theater to block all cellular telephone calls. **Note: The reader is advised to understand and obey all laws applicable to interference in telecommunication services. This device may be illegal in your country or area.**

How It Works

The circuit has two stages. One of them is formed by a high-frequency oscillator that runs in the VHF band. This circuit uses a 2N2218 transistor. The frequency is determined by L1 and the capacitance provided by the varicap diode.

The second stage is the modulation circuit, which acts on the varicap, thereby changing the frequency produced by the oscillator circuit within a broad band. This stage picks up the 60 Hz power line signal and passes it through a diode bridge, producing a 120 Hz dc pulsed voltage.

Through P1, this voltage is applied to the varicap diode, modulating the signal. In this way, the high-frequency signal changes its frequency spread within the FM band at a rate of 120 times per second as shown in Fig. 3.1.

When any receiver that is located near the transmitter is tuned to any point to the FM band, the noise signal will be received at this point 120 times per second, producing a strong 120 Hz hum.

Because the circuit is just intended to produce noise, it does not require any special layout.

Assembly

Figure 3.2 is a schematic diagram of the noise transmitter. The components are installed on a printed circuit board as shown in Fig. 3.3.

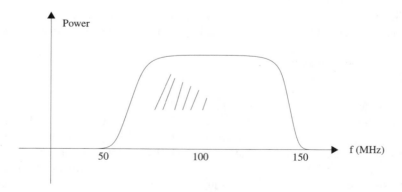

Figure 3.1 Band covered by the transmitter (typical).

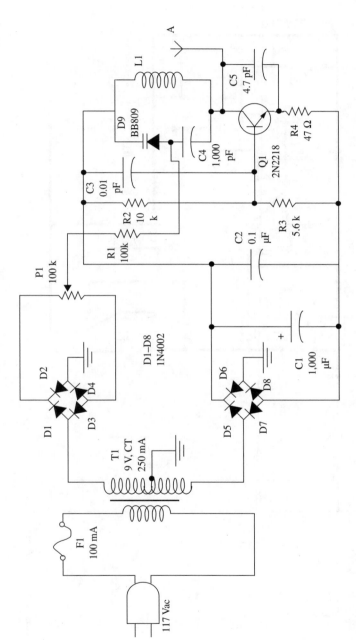

Figure 3.2 Noise transmitter.

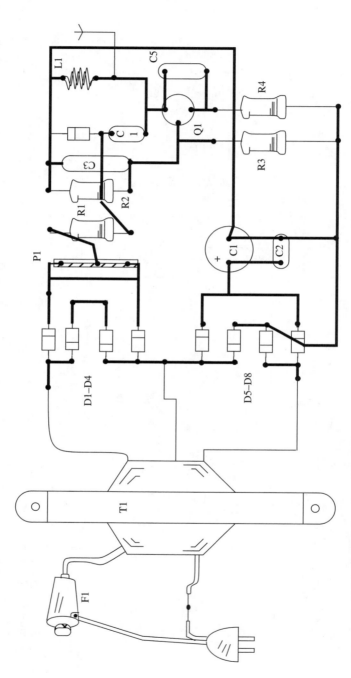

Figure 3.3   Printed circuit board used for Project 16.

The transformer consists of a 117 Vac primary winding and any 7.5 to 9 V CT secondary, with currents rated to 250 to 500 mA. The diodes are all 1N4002 or equivalents except D9. D9 is any common varicap such as the BB809 or equivalent. As explained in other projects that use varicaps, this component can be found in the tuning circuits of old touch-key TV sets and FM radios that are tuned by potentiometers.

Coil L1 is formed by three to five turns of AWG 18 to 22 enameled or plastic-covered rigid wire on a 1 cm dia. coreless form. You can use a pencil as the reference for winding this coil.

The antenna is a piece of rigid plastic-covered wire 20 to 40 in. long, or a common telescoping antenna.

All of the capacitors must be ceramic types except C1. C1 is an electrolytic capacitor rated to 16 V or more. The circuit can be easily installed in a plastic box.

Testing and Adjusting the Circuit

At a distance of 3 to 9 ft away from the transmitter, place an FM receiver tuned to a free point in the FM band or to any local radio station. Turning on the transmitter (by plugging it to the ac power line), open P1 until a strong hum is produced by the receiver. This is the only necessary adjustment. The transmitter is ready for use.

How to Use

Place the transmitter in any location where you suspect a spy transmitter may exist. The signal will cancel out the spy signal if it is within the operational range. (Of course, your friends also will not be able to tune to their favorite programs in the FM band. This suggests an interesting application for the device if someone in the office likes to listen to awful music programs....)

Parts List: Project 16

Semiconductors

Q1	2N2218 or equivalent high-frequency NPN silicon transistor
D1–D8	1N4002 or equivalent silicon rectifier diodes
D9	BB809 or equivalent variable capacitance diodes (see text)

Resistors (1/8 W, 5%)

R1	100,000 Ω—brown, black, yellow
R2	10,000 Ω—brown, black, orange
R3	5,600 Ω—green, blue, red
R4	47 Ω × 1 W—yellow, violet, black
P1	100,000 Ω—trimmer potentiometer

Parts List: Project 16 (continued)

Capacitors

C1	1,000 μF/16 WVDC electrolytic
C2	0.1 μF ceramic
C3	0.01 μF ceramic
C4	1,000 pF ceramic
C5	4.7 pF ceramic

Additional Parts and Materials

F1	200 mA fuse
T1	transformer, 117 Vac primary winding, 7.5 to 9 V × 150 mA secondary (see text)
L1	coil (see text)
A	antenna (see text)

Printed circuit board, plastic box, power cord, fuse holder, wires, solder, etc.

Project 17: Alert FM Transmitter

If the sensor wired to this transmitter is triggered on, an alert signal will be emitted that can be received by any FM transmitter placed at a distance up to 300 ft.

Features

- Power supply: 6 Vdc, 4 AA cells
- Operation frequency: 88 to 108 MHz
- Range: up to 300 ft (basic version)
- Alarm signal: audio tone between 100 and 2 kHz

This circuit can be used to monitor remote sensors such as alarm sensors, cage sensors, switches placed in doorways, etc. If the sensor is closed or opened (depending on how it is connected), it triggers an oscillator that modulates the high frequency signal emitted by the transmitter.

Several interesting applications are suggested, as follows:

- It can be used as a wireless car alarm. The transmitter will be activated, sending a signal to a remote FM receiver when someone tries to start the engine or open the door. The alert signal can be picked up by a bedside FM radio.
- For espionage, it can detect when an enemy is coming. A sensor can be placed in a strategic place to activate the transmitter when an intruder is detected.

- The transmitter can be used to send a signal when a cage door is opened by an escaping animal, or when the door of a trap is closed after an animal is captured. The receiver can be placed at a secure distance.
- A building can be protected against intruders by installing sensors in the doors and windows. Any intruder will trigger the circuit, sending an alert signal to your FM receiver.

How It Works

The basic idea behind this alarm is very simple. An audio oscillator using a unijunction transistor (UJT) is coupled to a high-frequency oscillator tuned to a free point in the FM band. The high-frequency oscillator is the transmitter, which is on the air at all times. However, the audio frequency oscillator triggers only when a sensor is activated.

The simplest sensor is a piece of thin wire placed between two nails fixed in a doorway as shown Fig. 3.4b. When the wire is connected in the circuit, the audio oscillator is off. But if the wire is cut (when the door or window is opened), R1 will bias the oscillator, turning it on.

Another sensor is shown in Fig. 3.4a. This is a microswitch, reed switch, or simple push button. When the sensor closes the circuit, the oscillator is biased on.

The circuit is powered from 4 AA or D cells for electrical autonomy. If it is used in a car, a voltage reducer can be used to power it from the 12 V battery.

Assembly

A schematic diagram of the Alert Transmitter is shown in Fig. 3.5. The components are placed on a printed circuit board as shown in Fig. 3.6.

The coil (L1) consists of four turns of AWG 18 to 22 enameled or plastic-covered rigid wire wound over a 1 cm dia. form. Use a pencil as reference for winding this coil.

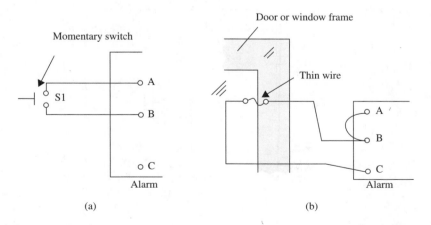

Figure 3.4   Simple sensors.

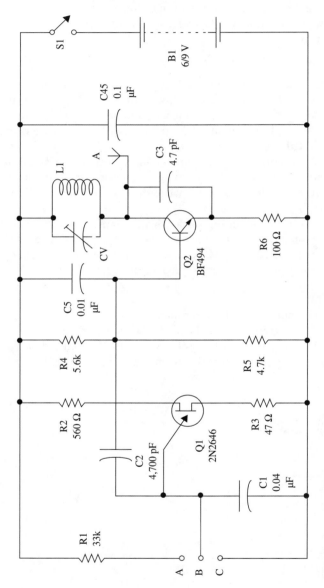

Figure 3.5   Alert FM transmitter.

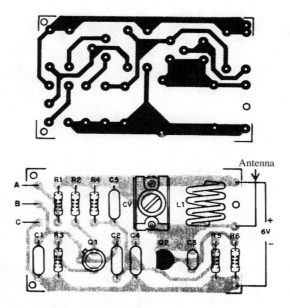

Figure 3.6 Printed circuit board used in Project 17.

Any trimmer capacitors with capacitances ranging from 2–20 to 4–40 pF can be used. Plastic or porcelain types are suitable for this project. The antenna is a piece of rigid wire 10 to 30 in. long or a telescoping antenna.

Q1 is a unijunction transistor (UJT). The reader must be careful when installing this transistor. If it is placed in inverted position, the circuit will not operate properly.

The original transistor for 6 V applications is the BF494 or an equivalent such as the BF495. But if the reader wants a more powerful circuit, it can be powered from 9 to 12 V supplies. In this case, replace Q1 with a 2N2218 or BD135, and replace R6 with a 47 Ω × 1 W resistor. The signals can be picked up at distances up to 900 ft with these modifications.

To power the 6 V circuit from a 12 V supply, you can use the voltage reducer shown in Fig. 3.7. The integrated circuit must be mounted on a small heat sink. The circuit can be plugged into the car's electrical system using the cigarette lighter socket.

Adjustment and Use

Place an FM receiver tuned to a free point in the FM band near the transmitter. Connect a piece of wire between points A and B, and turn on the transmitter power supply.

Tune the circuit with the trimmer capacitor (CV) until you have the strongest signal in the receiver. If you don't like the sound of the audio tone coming from the radio, you can change the value of R1. Values between 22,000 and 100,000 Ω

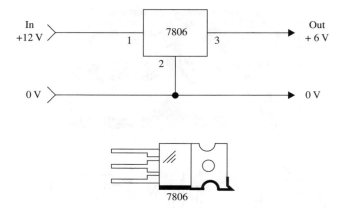

Figure 3.7 Powering the circuit from a 12 V power supply.

can be tested. You can also replace this resistor with a 100,000 Ω trimmer potentiometer in series with a 10,000 Ω resistor to adjust the tone more precisely.

When the piece of wire between points A and B is disconnected, the radio signal tuned to the receiver frequency continues, but the audio signal does not.

Figure 3.8 shows how the sensor can be wired to the circuit. You can use normally open (NO) sensors or normally closed (NC) ones, depending on how you
want the circuit to operate. The wires to the sensor can be very long (up to 300 ft)
without causing operational problems.

In Fig. 3.8, the circuit is arranged for use as a remote car alarm. When one of
the door-activated switches is closed, the courtesy lights are turned on, and the
circuit begins to send the alert signal to a remote FM receiver.

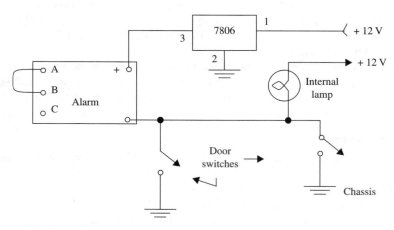

Figure 3.8 Installing the circuit in a car.

Parts List: Project 17

Semiconductors

Q1	2N2646 unijunction transistor (UJT)
Q2	BF494 or equivalent high-frequency, small-signal NPN transistor (see text)

Resistors (1/8 W, 5%)

R1	33,000 Ω—orange, orange, orange
R2	560 Ω—green, blue, brown
R3	47 Ω—yellow, violet, black
R4	5,600 Ω—green, blue, red
R5	4,700 Ω—yellow, violet, black
R6	100 Ω—brown, black, brown

Capacitors

C1	0.047 µF ceramic or metal film
C2	4,700 pF ceramic
C3	4.7 pF ceramic
C4	0.1 µF ceramic
C5	2,200 nF ceramic
CV	trimmer capacitor (see text)

Additional Parts and Materials

L1	coil (see text)
B1	6 V, four AA, C, or D cells or voltage reducer from the car's battery
A, B, and C	isolated terminals

Printed circuit board, battery holder, plastic box, sensors, antenna, wires, solder, etc.

Project 18: Wireless FM Event Monitor (Version 1)

This circuit is useful for the detection of a falling object, the opening or closing of a door, the passage of a human body, and similar events.

Features

- Power supply: 6 Vdc (four AA cells)
- Range: 150 ft (basic version)
- Frequency range: 88 to 108 MHz
- Sensors: LDR for the basic version

How can we monitor a door without being there? How can we detect remote phenomena in scientific experiments without using wires? The solution is this event monitor/transmitter. The device sends out a signal whenever something triggers a sensor.

Our basic version uses a light sensor (photoresistor, also known as a light-dependent resistor or LDR) to detect the presence of objects or persons when they interrupt a light beam or change the amount of light falling onto the sensor. For example, by placing a tube filled with a chemical solution between the sensor and a light source, it is possible to detect when the solution inside the tube changes from transparent to cloudy. In this way, we can monitor chemical processes from a distance, with no physical connection between the monitor and FM transmitter.

Other interesting applications for this project are as follows:

- A secret agent, detective, or spy can use the circuit to detect an intruder.
- A scientific researcher can follow experiments at a distance without leaving the laboratory.
- Via a remote receiver, a shopkeeper can detect when a customer enters the shop.

How It Works

A block diagram that illustrates the operating principle of this project is shown in Fig. 3.9. The circuit is formed by a transmitter that is modulated by spaced

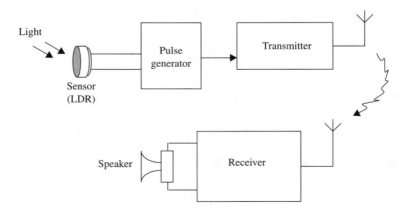

Figure 3.9 Block diagram representing the systems.

"beeps." The beeps are generated by a low-frequency oscillator using the 555 integrated circuit.

The separation between the beeps is controlled by the sensor. Therefore, if the amount of light falling onto the sensor changes, so does the beep rate.

The recommended sensor is an LDR or photoresistor, a device whose electrical resistance depends on the amount of light falling on a sensitive surface. In the dark, the device has a very high electrical resistance—in the range of several megohms. When exposed to light, the resistance falls to several hundred ohms or less.

This sensor can be mounted inside an opaque tube with a convergent lens placed in front of it. In this way, it is possible to increase its sensitivity and directivity as shown in Fig. 3.10.

The LDR is wired to the base of a transistor that controls the frequency produced by an astable 555. In the 555 oscillator, the frequency is governed by C1, resistors R2 and R3, and the resistance between the emitter and collector of Q1, depending on the amount of light that falls onto the sensor. Therefore, a change in the amount of light also changes the time constant in the oscillator circuit and thereby the frequency. The trimmer potentiometer is used to adjust the central frequency produced by the oscillator as appropriate for the amount of light used to control the circuit.

In normal operation, the circuit is adjusted to produce pulses separated by wide time intervals. But, at the instant in which a person passes in front of the sensor and thereby reduces the amount of light falling onto it, the frequency increases, shortening the interval between the beeps.

Figure 3.11 graphically illustrates how the beeps change in frequency rate by a ratio of about 10:1.

An LED wired to the output of the oscillator acts as a monitor. After this point in the circuit, the pulse is inverted and used to control an audio oscillator that uses another 555 IC, as shown in Fig. 3.12.

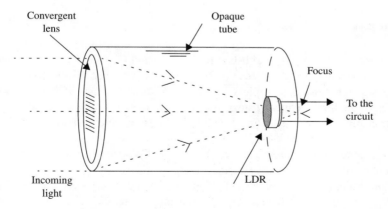

Figure 3.10   Increasing circuit sensitivity.

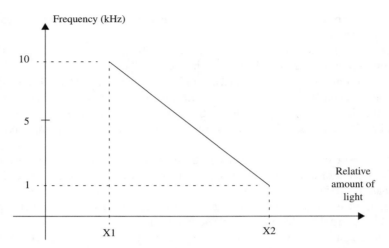

Figure 3.11 Response of the circuit to the amount of light.

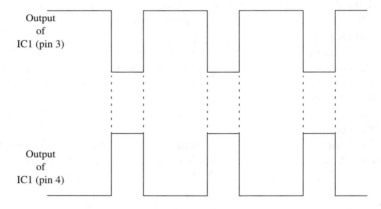

Figure 3.12 Wave shapes at two points in the circuit.

The beep frequency is established by R7, R8, and C2. The reader can change these components to alter the tone (C2 in particular, between 10 and 100 nF).

The separated bits can be picked up from the circuit output (IC2, pin 3) to be applied to the high-frequency stage. The high-frequency output stage, which acts as a transmitter, is formed using a BF494 in the same configuration as for many other projects described in this book.

The frequency is determined by L1 and CV. CV must be adjusted to a free point in the FM band.

If the reader requires more power in this stage, other circuits can be used. Elsewhere in this book, the reader can find transmitter circuits that can send signals over distances up to several miles. The simplest modification to increase the power is to replace transistor Q3 with a 2N2218 and the power supply by 9 or 12 V batteries.

Assembly

The circuit diagram for the Event Transmitter is shown in Fig. 3.13. The components are placed on a universal printed circuit board as shown in Fig. 3.14.

Notice that this configuration can be used to test the circuit on a solderless board. Because this is a high-frequency circuit, it is important to keep all connections as short as possible.

Any LDR can be used in this circuit, as can other resistive sensors such as NTCs or PTCs. You can find an LDR in any old TV set that contains an automatic brightness control. The LDR can be wired to the circuit via a long cable. Placing the LDR at a distance from the transmitter allows optimal transmitter placement for the best transmission characteristics.

The trimmer capacitor can range from 2–20 to 4–40 pF, and any type, porcelain or plastic, can be used. The coil is formed by four turns of AWG 18 to 22 enameled or plastic-covered rigid wire on a 1 cm dia. form. Use a pencil as reference for winding the coil. You can alter this coil to transmit signals in other bands. With six or seven turns, it is possible to transmit signals in the low VHF band between 40 and 70 MHz. The antenna can be a piece of plastic-covered rigid wire 10 to 30 in. long or a telescoping antenna.

Adjustment and Use

Place near the transmitter any FM receiver tuned to a free point of the FM band. Power on the transmitter and adjust CV to tune the signal. A series of beeps will be reproduced by the receiver's loudspeaker.

Adjust P1 to change the beep rate. When you pass your hand in front of the sensor (LDR), the beep rate should change. Experiment to find the correct P1 adjustment for the application you have in mind. If necessary, install the LDR in an opaque tube with a lens to increase sensitivity.

To use the device, aim the sensor to pick up light changes caused by the event to be detected. Tune the receiver and pay attention to frequency changes. Any frequency change means that something happened to change the amount of light falling onto the LDR.

Figure 3.15 shows the circuit used to detect when someone enters a room. The light source is not always necessary; under some circumstances, ambient light may be sufficient to act on the sensor.

How to Make a Small-Capacitance Capacitor

If you have trouble finding the small capacitor C5, a home-made replacement can be fabricated. This is not a critical component, and values between 2.2 and 5 pF normally can be used in these projects. It is not necessary to match the exact dimensions of the commercial parts. Figure 3.16 shows how such a capacitor can be made using interwoven, plastic-covered, rigid wires.

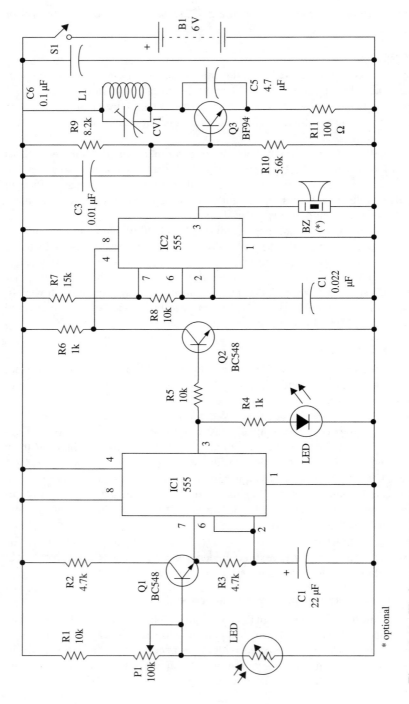

Figure 3.13 Wireless event transmitter.

* optional

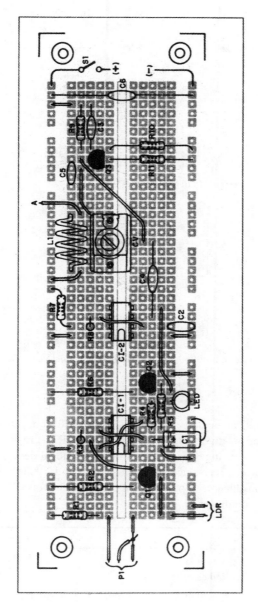

Figure 3.14   The circuit can be mounted on a solderless board as shown above.

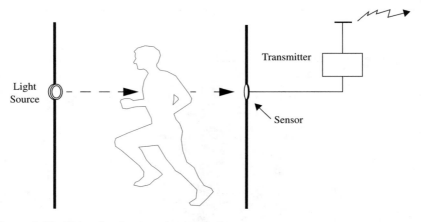

Figure 3.15   Using the circuit to detect people.

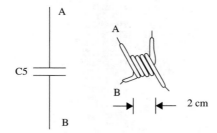

Figure 3.16   Using wire to build a low-capacitance capacitor.

Other Sensors

Temperature changes can be detected using sensors such as negative temperature coefficient (NTC) or positive temperature coefficient (PTC) resistors. The resistance of these devices changes with temperature. The ideal device to be used in this circuit must have a resistance between 10,000 and 100,000 Ω at ambient temperatures. A transducer can be wired into the circuit for self-monitoring as shown in the schematic diagram.

Parts List: Project 18	
Semiconductors	
IC1, IC2	555 integrated circuits, timer
Q1, Q2	BC548 or equivalent general purpose NPN silicon transistors
Q3	BF494 or 2N2218 high-frequency NPN silicon transistor (see text)
LED1	common red LED

Parts List: Project 18 (continued)

Resistors (1/8 W, 5%)

R1, R5, R8	10,000 Ω—brown, black, orange
R2, R3	4,700 Ω—yellow, violet, red
R4, R6	1,000 Ω—brown, black, red
R7	15,000 Ω—brown, green, orange
R8	8,200 Ω—gray, red, red
R10	5,600 Ω—green, blue, red
R11	100 Ω—brown, black, brown
P1	100,000 Ω—trimmer or common potentiometer (linear or log)

Capacitors

C1	22 µF/12 WVDC, electrolytic
C2, C4	0.022 µF, ceramic or metal film
C3	0,01 µF, ceramic
C5	5.6 or 4.7 pF, ceramic
C6	0.1 µF, ceramic
CV	trimmer capacitor (see text)

Additional Parts and Materials

LDR	Any light dependent resistor (LDR), photoresistor, or CdS cell (see text)
BZ	piezoelectric transducer (see text)
L1	coil (see text)
S1	SPST toggle or slide switch
B1	6 V, four AA cells or 12 V supply when using the 2N2218 (see text)

Printed circuit board, battery holder, plastic box, hardware for the LDR (tube, lens, etc.), wires, solder, etc.

Project 19: Wireless Event FM Transmitter (Version 2)

Some improvements were added to the previous wireless event FM transmitter to produce a second version. These improvements change the circuit's performance and range of functions.

Features

- Power supply voltage: 3 to 12 Vdc
- Range: 150 to 1,500 ft (according to the power supply, see text)
- Operating frequency range: 88 to 108 MHz

As with the previous project, this circuit is intended for wireless monitoring of remote events using a common FM receiver. This circuit is different from the first version in that, rather than sending beeps at a repetition rate that varies with the resistance of the sensor, this version sends constant beeps when the sensor is triggered.

Many types of sensors can be used with this circuit, as suggested by the following applications:

- *Water sensor.* The circuit is triggered when water falls on a sensor. There are two ways to trigger the circuit: when the water level drops below a predetermined value or when the water level rises above a certain value. The circuit can be used to control the level in a reservoir or to detect rain. For this application, the sensor is formed by two bare wires separated at a distance of one inch, or two metal screens with a salted tissue between them.
- *Position sensor.* The movement of objects, doors or windows closing or opening, and other displacements of people, objects, and animals can be detected with an appropriate sensor. The recommended sensor can be a blade magnetic switch, microswitch, reed switch, or any other type.
- *Light sensor.* For this application, the sensor can be a light dependent resistor (LDR) or a photoresistor (or CdS cell, as it is also called). The circuit can detect when a light is turned on or off, when a person enters in a room, or when a substance changes its transparency due to a chemical reaction.
- *Temperature sensor.* NTCs, PTCs, or diodes can be used to detect temperature changes. The circuit will send a signal when the monitored temperature falls below or rises above a preset value.
- *Other sensors.* As appropriate to the application, the reader will discover that other sensors can be used, some of which can be home made. For instance, pendulum sensors can detect when an object is shaken. It is up to the reader's imagination to create the necessary sensor for an application.

How It Works

Figure 3.17 shows a block diagram of this circuit. With reference to this diagram, we will explain how the circuit works.

The first block is a monostable multivibrator activated by the sensor. The objective of this circuit is to produce a constant-duration pulse, independent of the time in which the sensor is activated. This kind of action is important for applications where the sensor is activated by short time intervals.

The pulse duration can be adjusted to values between several seconds and five minutes with the components values shown in the diagram. The time values in this range are adjusted by P1. If the reader wants longer time intervals, both P1 and C1 values can be increased. P1 values up to 1,500,000 Ω and to C1 values up to 2,200 μF are permitted, extending the maximum time interval up to an hour.

The integrated circuit IC1, which acts as the monostable, is triggered when pin 2 is driven to a low level. This means that, to trigger the circuit, point X1 must be

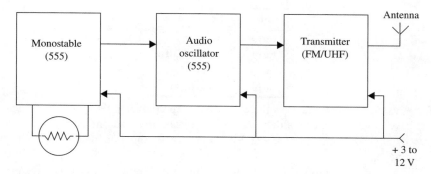

Figure 3.17 Block diagram of Project 19.

connected to ground, or at least to a voltage corresponding to one-third or less of the power supply voltage.

Note that Rx keeps this input (pin 2) at the high level, and so the sensor must form a voltage divider with this component. Therefore, to trigger the circuit on, the value of Rx must be carefully chosen to put the voltage at pin 2 below one-third of the power supply voltage at the triggering point.

A simple rule to adopt when choosing Rx is that this component must have twice the sensor resistance value when the circuit is to be triggered. For instance, if the sensor is an LDR with a resistance range between 10,000 and 1,000,000 Ω in the proper light range, a resistor of 100,000 Ω can be used, assuming that the desired level to trigger the circuit will occur when the LDR presents a 50,000 Ω resistance.

If low-level light sources are used, the value of Rx must be increased to 470,000 Ω. This means that the circuit will be triggered on when the LDR resistance falls to approximately 220,000 Ω.

For easy adjustment, the reader can add a potentiometer in series with Rx. A 1,000,000 Ω component is suitable.

The described monostable stage controls a second block, which also uses a 555 integrated circuit. But, in this case, the IC is wired as an audio oscillator.

The audio oscillator has its frequency determined basically by C3 and adjusted by P2. The tone can be adjusted to frequencies in the range between 500 and 5,000 Hz. The circuit is off until the output of the first block goes to the high level. This occurs only when the first block is triggered on. Therefore, an audio tone is produced and applied to the input of a high-frequency stage, which acts as a transmitter.

The transmitter is formed by one transistor in the common configuration found in many other projects described in this book. The produced frequency is in the FM range, but if the reader has a VHF receiver, it is possible to find a less congested band in that range for operation of the circuit. By changing the coil, it is possible to alter the frequency as shown in the following table.

Coil	Frequency
3 turns	108 to 115 MHz
5 turns	70 to 88 MHz
6 turns	50 to 70 MHz
7 turns	40 to 50 MHz

These values are not exact, as the components (particularly the trimmer capacitor) have large tolerances in their values. Therefore, the reader must find the correct number of turns experimentally according to the intended transmission frequency.

The transistor and the power supply determine the output power and therefore the range. Using a BF494 transistor, the circuit can be powered from 3 to 6 V power supplies; using a 2N2218 device, the circuit can be powered from a 9 or 12 V supply.

The following table gives the components' values for the several usable voltages to power the circuit and the range achieved in each case.

Supply voltage	R4	R5	R6	Range (ft)
3 V	4.7k	5.6k	47	150
6 V	5.6k	8.2k	47	450
9 V	6.8k	8.2k	100	900
12 V	8.2k	10.0k	100	1500

The antenna also affects performance, and pieces of rigid plastic-covered wire between 10 and 40 in. in length can be used for this task. It is important to find the best place to connect the antenna to the coil, as described for many transmitters in this book.

Assembly

Figure 3.18 shows a schematic diagram of this transmitter. Notice that some components have values that depend on the power supply voltage and the output transistor. It is also important to observe that, for the 3 V version, a bipolar 555 will not operate as expected. This component must be replaced by a CMOS 555 or a TLC7555 as specified.

The components are placed on a printed circuit board as shown in Fig. 3.19. For the 12 V version, the transistor must have a small heat sink.

The specifications for resistors and capacitors are given in the parts list. Note that there are some ceramic capacitors that **must not** be replaced by other types.

The coil is formed by four turns of AWG 18 to 22 enameled wire or common rigid plastic-covered wire wrapped around a pencil as a reference (diameter = 1 cm). The tap to use for the antenna connection must be found experimentally.

Any trimmer capacitor with capacitance ranges between 2–20 and 4–40 pF can be used. Porcelain or plastic types are suitable. The reader only has to observe the

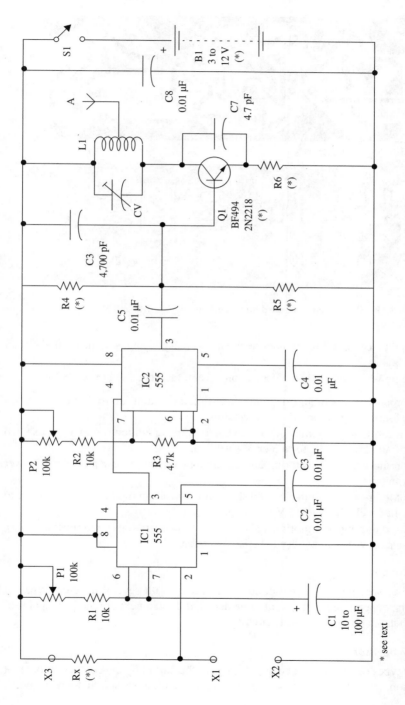

Figure 3.18 Wireless event transmitter (version 2).

* see text

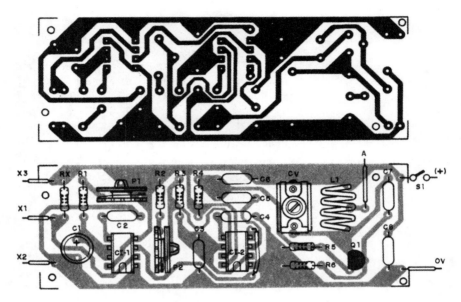

Figure 3.19   Printed circuit board for the circuit shown in Fig. 3.18.

terminal placement if it becomes necessary to make changes in the printed circuit board layout.

The power supply choice depends on several factors as suggested below:

1. For short-range monitoring and short-time operation (up to 20 min), use two or four AA cells for a 3 or 6 V version.
2. For short-range monitoring and extended-time operation (up to 3 hr), use two or four C or D cells for a 3 or 6 V version.
3. For short-range monitoring and continuous operation, use a 3 or 6 V power supply.
4. For long-range monitoring and short-time operation (up to 20 min), use six to eight C cells for 9 to 12 V versions.
5. For long-range monitoring and extended-time operation (up to several hours), use a car or nicad battery for a 12 V version.

The Sensors

As described previously, the circuit can operate with several different types of sensors, depending on the event to be detected. Listed next are the principal sensors that can be used with this circuit.

Contact Sensors

Many types of sensors can be used to trigger the circuit by the contact or touch of a person, animal, or object. A push button, a pendulum sensor, and a reed switch are samples of sensors that can be used for this task. Some of these sensors are shown in Fig. 3.20. The recommended range of values for Rx when using these sensors is between 4,700 and 47,000 Ω.

Such sensors can be used to detect when a cage door is opened or closed by an animal or when objects are moved. The sensor can be mounted in the same box used to install the transmitter or wired to it by a long, flexible wire. It is not necessary to use shielded cables.

Water Sensors

The appropriate sensors for water detection are shown in Fig. 3.21. The sensors as wired in two cases (Fig. 3.21a and b) are indicated to detect changes in water

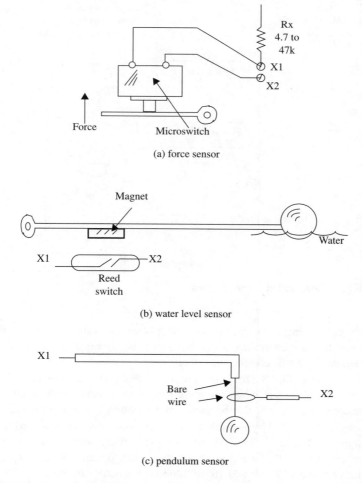

(a) force sensor

(b) water level sensor

(c) pendulum sensor

Figure 3.20   Some sensors that are suitable for this project.

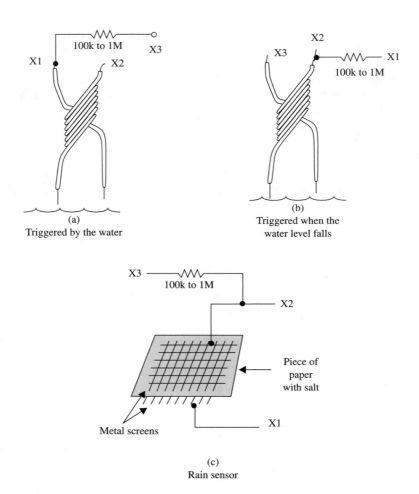

Figure 3.21 Water and moisture sensors.

level. This configuration can be used to detect when the water level in a reservoir falls or rises to a predetermined point. Note that the circuit is triggered when the water touches the wires, closing the circuit.

The third sensor (Fig. 3.21c) is used to detect any presence of water. This sensor can be used to detect when a rain begins or to detect plumbing leaks. The two-wire configuration can also be used to detect when a flower pot needs water.

The sensitivity offered by circuits that use this sensor can be increased with the circuit shown in Fig. 3.22. This circuit uses a sensitivity control, and the capacitor value depends on the speed at which the events occur. You can experiment with values between 470 nF and 10 μF to find out which is appropriate for a particular application.

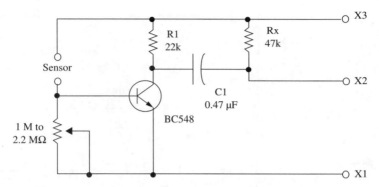

Figure 3.22   High-speed circuit used with resistive sensors.

One suggested application for this circuit is detecting a wave when it passes by the sensor. The circuit can be used as a swimming pool alarm, detecting when someone falls accidentally into the water.

Light Sensors

The best sensor for this application, because of its sensitivity and low cost, is the LDR or light dependent resistor (CdS cell). The circuit used to wire this sensor to the event transmitter is shown in Fig. 3.23.

Rx is replaced by the potentiometer or the LDR, depending on the version. In the first circuit, the sensor triggers on the transmitter when light falls onto its sensitive surface. The potentiometer adjusts sensitivity as appropriate to the ambient light level.

The second version is triggered when the light falling onto the sensor is cut off or reduced. Here, too, sensitivity is adjusted using the potentiometer.

If the light changes to be detected are very short in duration, the circuits shown in Fig. 3.24 are more appropriate. One of the circuits triggers on the transmitter when a short pulse of light falls onto the sensor. The other circuit triggers the transmitter when a short light outage occurs.

In all cases, directivity and sensitivity can be increased if the LDR is mounted inside an opaque tube with a convergent lens placed in its opening.

Temperature Sensors

A negative temperature coefficient (NTC) or positive temperature coefficient (PTC) resistor can be used to detect temperature changes. The circuits are basically the same as the ones used with other resistive sensors, such as the LDRs. Figure 3.25 shows some circuits that can be used with these sensors.

Another sensor that can be used to detect temperature changes is a reverse-biased silicon diode. A transistor must be used to amplify the small temperature-dependent current that flows by this device. Note that circuits using NTCs act in the opposite manner of PTC-based circuits.

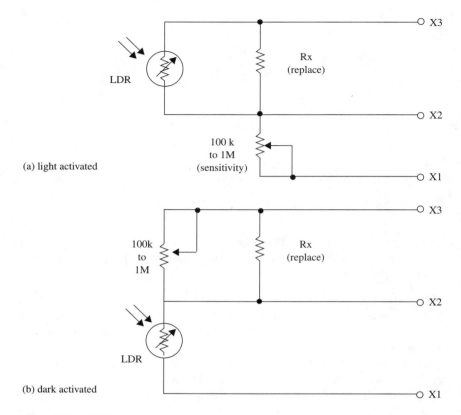

Figure 3.23   Light sensors.

Other Sensors

A piece of conductive foam and two metal plates can be used to make a pressure sensor as shown in Fig. 3.26. An appropriate source of this foam is the packing material used to transport sensitive CMOS integrated circuits, dissipating static discharges that can destroy them.

Another sensor is shown in the same figure, using an LDR and a light source (a small incandescent lamp). This sensor can be used to detect when the transparency of a substance changes due to a chemical reaction.

Conclusion

The reader who makes scientific field experiments can find many uses for this project. If you install the transmitter in a plastic box and keep a stock of sensors on hand, you will be ready to perform a range of experiments using remote sensors whenever the need arises.

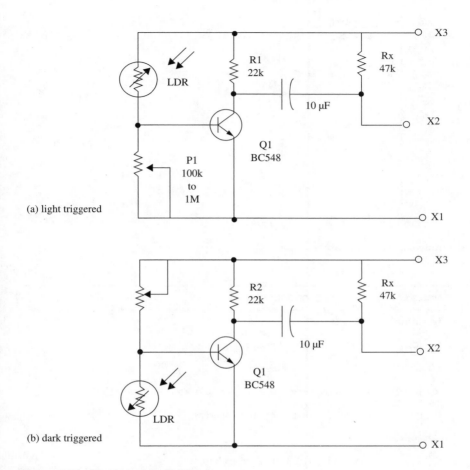

(a) light triggered

(b) dark triggered

Figure 3.24   More sensitive light sensors.

Adjustments

To use the circuit, the main required adjustment is performed using CV, with which you tune the circuit to a free point in the FM band. The other adjustments depend on the sensor and are implemented to find the most appropriate sensitivity for the application.

Parts List: Project 19

Semiconductors

IC1, IC2	555 integrated circuit, timer
Q1	BF494 or 2N2218 high-frequency NPN silicon transistor (see text)

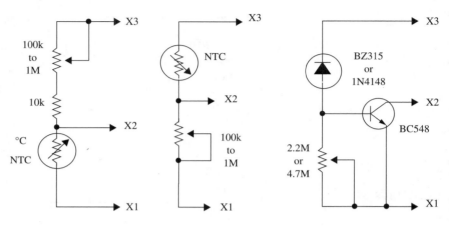

Figure 3.25   Temperature sensors.

Parts List: Project 19 (continued)

Resistors (1/8 W, 5%)

R1, R2   10,000 Ω—brown, black, orange

R3   4,700 Ω—yellow, violet, red

R4, R5, R6   values according supply voltage and transistor (see table)

P1, P2   100,000 Ω—trimmer potentiometers

Capacitors

C1   10 to 100 µF/12 WVDC, electrolytic

C2, C4, C5   0.01 µF, ceramic

C3, C8   0.1 µF, ceramic

C6   4,700 pF, ceramic

C7   4.7 pF, ceramic

Additional Parts and Materials

L1   coil (see text)

CV   2–20 to 4–40 pF trimmer capacitor (see text)

S1   SPST toggle or slide switch

B1   3 to 12 V power supply (see table and text)

Printed circuit board, plastic box, battery holder, sensors, antenna, wires, etc.

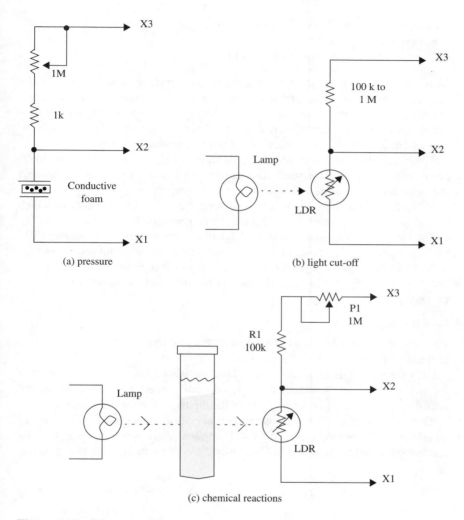

Figure 3.26   Other sensors.

Project 20: Wireless FM Alarm

With this device, you can send an alarm signal to a nearby FM receiver when your car is invaded or when any protected object is removed from its place.

Features

- Power supply voltage: 12 V (transmitter), 6 V (receiver interface)
- Frequency range: 88 to 108 MHz

- Modulation signal: 2,400 Hz (approx.)
- Range: 150 ft (typical)
- Detecting filter: phase locked loop (PLL)

Many people must leave their cars parked on the street, because they have more cars than garage slots, or perhaps no garage at all. Meanwhile, the common alarms that work through the horn are not highly desirable, as they can be triggered at inappropriate times, such as during a storm or late at night, when it is inconvenient to turn them off (unless they employ a remote control).

The circuit described here acts as an alarm but sends the signal to a remote receiver without the use of wires. This is important when it is impossible to connect the protected object as described in the previous project.

This system has other applications as well. Scientific experiments in which sensors are used to detect specific events can be performed using this circuit.

The circuit is composed of a transmitter that sends a signal to a common FM receiver tuned to a free point in the FM band. The receiver is plugged into an interface that controls a relay. Any powerful horn or other annunciator can be triggered by the relay. A PLL filter is used in the interface circuit to avoid erratic triggering.

How It Works

The system is formed basically by a transmitter and a receiver. A block diagram of these units is shown in Fig. 3.27.

Let's begin with the transmitter unit. This unit can be installed in a car or other location that is to be protected. When any door is opened, the sensors act on the circuit by momentarily grounding one of the inputs of a monostable circuit.

This circuit uses a 555 device, and its output goes to the high level for a time interval that is determined by R2 and C4. Using the components shown in the diagram, the interval is about 5 min. The reader, according to the application he has

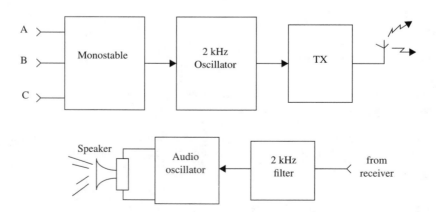

Figure 3.27 Block diagram for the wireless FM alarm.

in mind, can change these components within a wide range of values. R2 can be increased up to 2,200,000 Ω and C4 to up to 1,500 μF for time intervals up to 1 hr.

The monostable multivibrator triggers on an oscillator using another 555. The frequency produced by this circuit is determined by R3, R4, and C5. With the recommended components, the circuit runs at 2,400 Hz. This signal is used to modulate the high-frequency stage, which acts as a transmitter.

As the transmitter is on the air all the time, it is easy to locate its signal with the receiver when adjusting or installing the system. Only when the monostable is triggered by the sensor is the oscillator enabled, producing the low-frequency signal.

For the basic version, a BF494 device is recommended. But if you need more power, you can replace this transistor with a 2N2218 and change the value of R7. This component should be increased to 56 Ω.

Keep in mind that the range of the signal can be affected by obstacles between the transmitter and receiver. This is illustrated in Fig. 3.28.

The signals are received by a common FM radio. The reader can use a standard bedside clock radio, which is powered from the ac line.

The received signal is connected to the alarm interface from the speaker terminals or from the earphone output. If you use the loudspeaker terminals, it is important to install a closed-circuit jack as shown in Fig. 3.29. This arrangement disconnects the speaker when the interface is plugged into the jack and reconnects it when the interface is unplugged and the alarm is not in operation.

The audio signal picked up from the loudspeaker is applied to a sensitive PLL filter. This filter uses an NE/LM567 integrated circuit as its core. When the tone produced by the alarm transmitter is recognized by this filter, its output goes to the low level, biasing transistor Q1, which is the BC558 in the receiver. At the same time, an LED glows, indicating that the PLL has tuned in the signal.

The transistor controls an audio oscillator formed by transistors Q2 and Q3. This oscillator drives a loudspeaker to produce a tone that is determined by R7

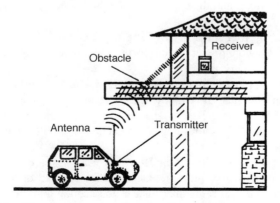

Figure 3.28 Metallic structures can affect signal propagation.

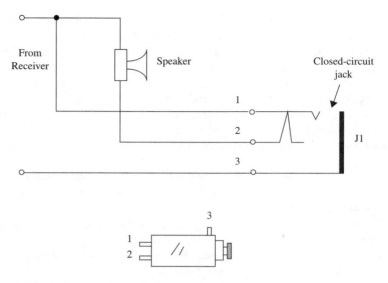

Figure 3.29   Adapting a closed-circuit jack for use in an FM receiver.

and C6. These components can be changed if the reader wants a different tone for the alarm. Values between 33,000 Ω and 100,000 Ω for R7 are recommended.

As the loudspeaker produces sounds only when the oscillator is activated, complete silence exists in the standby condition. The circuit is powered from the ac power line so that it will be on when the alarm is activated.

Assembly

A schematic diagram of the transmitter is shown in Fig. 3.30. The components are placed on a printed circuit board as shown in Fig. 3.31.

All component values are specified in the parts list. The electrolytics are rated to 12 WVDC or more, and the capacitor in the high-frequency stage must be ceramic.

L1 is formed by four turns of AWG 18 to 22 enameled or common plastic-covered rigid wire wound over a 1 cm form. Use a pencil as reference for winding this coil.

The antenna is a piece of plastic-covered rigid wire 10 to 30 in. long. This antenna must be placed near the window if the circuit is used in a car, as the metallic parts can act as obstacles to the radio signals and degrade transmitter performance. Figure 3.32 shows how the antenna can be positioned if the circuit is used in a car.

For the basic version, a BF494 or BF495 is used in the output stage. To get more power, the transistor can be replaced with a 2N2218 or a BD135. In this case, use a heat sink. But remember that, if you use the high-power version, the current drain will also be high, and the car battery can run down in a few hours.

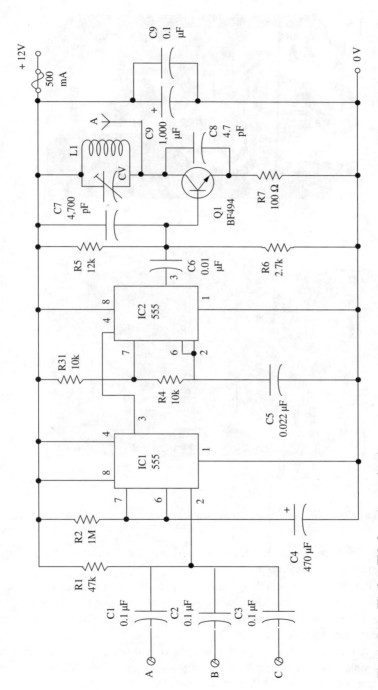

Figure 3.30   Wireless FM alarm (transmitter).

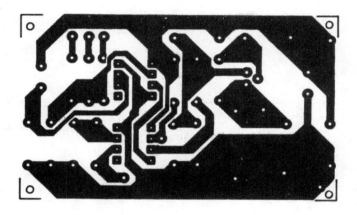

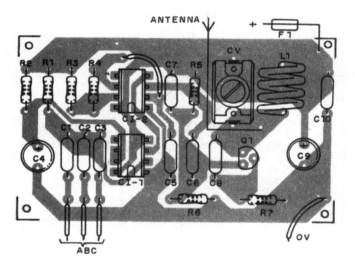

Figure 3.31   Printed circuit board used for Project 20.

Therefore, the high-power unit should be used only for short time intervals (up to one hour) or with an external power supply.

All of the components can be installed in a plastic box. If the antenna is placed at a distance from the unit, a shielded cable must be used to make the connection.

Figure 3.33 shows the schematic diagram of the receiver. The receiver components are mounted on a printed circuit board as shown in Fig. 3.34.

As the unit does not drain high currents, the IC-2 does not need to be installed on a heat sink. The transformer has a primary winding rated to 117 Vac and a secondary winding rated to 9 V CT with currents of 250 mA or more.

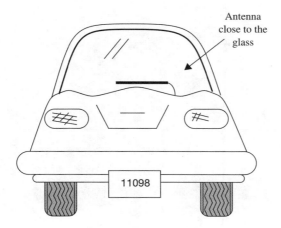

Figure 3.32   Antenna close to the glass.

All of the electrolytic capacitors are rated to 16 WVDC except C8, which is rated to 25 WVDC. A small loudspeaker (2 in.) can be used, as the circuit is placed near the person who monitors it, and the tone produced is loud enough to awake someone.

Plug the unit into the radio using a cable with a plug at the end, or install a switch to turn the circuit on and off as desired.

Adjustment and Use

First, plug the transmitter into a 12 V power supply. Tune any FM receiver to a free point of the FM band and place it near the transmitter. Adjust CV in the transmitter until you pick up the signal. If points A, B, and C are not wired, the received signal sounds like a whistle, covering the background noise that can be heard between stations.

After this, touch a wire between A, B, or C and the ground. The circuit will be triggered on, producing a tone in the receiver's loudspeaker. This sound will be produced for a time interval determined by the monostable components as described before.

Now, we move to the receiver adjustments with the interface. First, we plug the interface into the receiver, powering it on. Next, we trigger the transmitter on by momentarily touching a piece of wire to A, B, or C and the ground. While the tone is being produced by the circuit, adjust P1 in the receiver to tune in this tone. The LED will glow when the circuit is tuned.

After this, test the performance by taking either the transmitter or the receiver away to a greater distance and tweak the adjustments. When adjusted for best performance, you can install the circuit in the car or another place to be monitored. The sensor should be wired between points A, B, and C and ground so that it will trigger the circuit when closed.

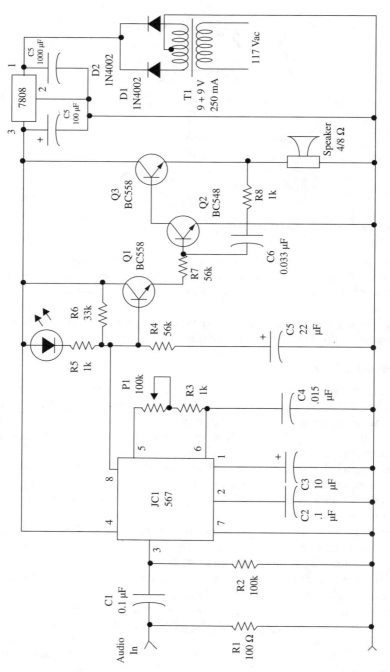

Figure 3.33 Wireless FM alarm (receiver).

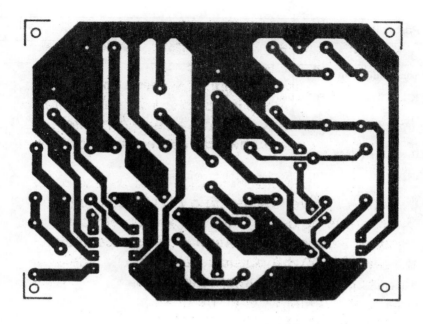

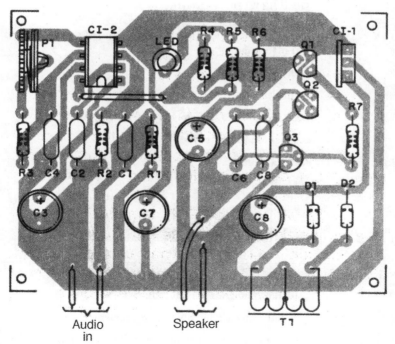

Figure 3.34 Printed circuit board for the receiver.

Parts List: Project 20

(a) Transmitter

Semiconductors

IC1, IC2	555 integrated circuits, timer
Q1	BF494 or 2N2218 high-frequency NPN silicon transistor (see text)

Resistors (1/8 W, 5%)

R1, R3, R4	10,000 Ω—brown, black, orange
R2	1,000 Ω—brown, black, red
R5	12,000 Ω—brown, red, orange
R6	82,000 Ω—gray, red, orange
R7	100 Ω—brown, black, brown

Capacitors

C1, C2, C3, C10	0.1 μF ceramic
C4	470 μF/12 WVDC electrolytic
C5	0.022 μF ceramic
C6	0.01 μF ceramic
C7	4,700 pF ceramic
C8	4.7 pF ceramic
C9	1,000 μF/16 WVDC electrolytic
CV	trimmer capacitor (see text) (2–20 to 4–40 pF)

Additional Parts and Materials

L1	coil (see text)
F1	500 mA - fuse and holder

Printed circuit board, plastic box, antenna, sensors, wires, solder, etc.

(b) Receiver

Semiconductors

IC1	LM/NE567 integrated circuit, PLL
IC2	7806 integrated circuit, 6 V voltage regulator
Q1, Q3	BC558 or equivalent general-purpose PNP silicon transistors
Q2	BC548 or equivalent general-purpose NPN silicon transistor

Parts List: Project 20 (continued)

D1, D2	1N4002 or equivalent silicon rectifier diodes
LED	Common red LED

Resistors (1/8 W, 5%)

R1	100 Ω—brown, black, brown
R2	100,000 Ω—brown, black, yellow
R3, R5, R6	1,000 Ω—brown, black, red
R4, R7	56,000 Ω—green, blue, orange
R6	33,000 Ω—orange, orange, orange
P1	100,000 Ω—trimmer potentiometer

Capacitors

C1, C2	0.1 µF, ceramic or metal film
C3	10 µF/12 WVDC, electrolytic
C4	0.015 µF, ceramic or metal film
C5	22 µF/12 WVDC, electrolytic
C6	0.033 µF or 0.047 µF, ceramic or metal film
C7	100 µF/12 WVDC, electrolytic
C8	1,000 µF/16 WVDC, electrolytic

Additional Parts and Materials

SPKR	4 or 8 Ω × 2 in. loudspeaker
T1	transformer, 117 Vac primary and 9 V CT secondary rated 250 mA or more

Printed circuit board, plastic box, power cord, jack and plug, wires, solder, etc.

Project 21: FM Beep Emitter

This circuit can be used in radio signaling or as a position locator when installed in a car or other object.

Features

- Power supply voltage: 6 V (four AA, C, or D cells)
- Range: 150 to 600 ft
- Frequency range: VHF and FM

There are many uses for radio signaling transmitters. One of the applications is to locate persons or objects. If the circuit is installed inside an object, this object can be found with the aid of a receiver equipped with a directional antenna.

An interesting game that can be played using this transmitter is called "fox hunt" by radio amateurs. The *fox* is the transmitter that is hidden somewhere, and the *hunters* are the radio amateurs equipped with receivers. The hunters must tune in the signal transmitted by the fox and locate it as quickly as possible. The winner is the first hunter to find the transmitter.

The reader can use this device to play the same game at school or while camping, with the advantage that anyone who has a receiver can tune in the signals. Any FM portable radio, Walkman®, or even a car radio can be used to chase the fox.

The basic circuit is powered from a 6 V supply (four AA cells) and can send signals to distances up to 600 ft over an open field. However, the circuit can be altered to use a 2N2218 and a 9 to 12 V power supply, thereby increasing the range.

How It Works

The high-frequency signals, up to 150 MHz, are produced by a transistor oscillator where the frequency is determined by L1 and CV. The feedback that keeps the oscillator running is provided by C5.

This capacitor allows us to produce higher frequencies with this circuit. Therefore, for the FM range, we use a 4.7 to 5.6 pF capacitor, but this value must be reduced if we intend to reach higher frequencies. Values between 1 and 2.2 pF are recommended for the range between 108 and 150 MHz. For the range between 50 and 80 MHz, this capacitor must be increased to 10 to 22 pF.

The beeps used to modulate the high-frequency carrier are produced by a 4093 CMOS IC. This integrated circuit is formed by four NAND Schmitt gates that can be used as oscillators, buffers, and/or inverters.

One of the four gates (pins 1 to 3) is used as a low-frequency oscillator to produce the beep rate. This circuit is designed to produce signals of about 0.5 Hz, as determined by R1 and C1, but these component values can be changed. The reader can vary R1 in the range between 100,000 and 4,700,000 Ω.

The beep tones are determined by a second oscillator that is wired using a second gate (pins 4 to 6). The frequency is determined by C2 and R2. Also, in this case, resistor R2 can be varied within a wide range to change the tone. The second oscillator is controlled by the first using pin 5.

The modulated signal that consists of a series of beeps is applied to the third gate. This gate is wired as a buffer/inverter to drive the transmitter.

The current drain depends on the RF transmitter stage, but for applications where it is kept in operation for one or two hours, four AA cells will provide sufficient power.

Assembly

Figure 3.35 shows the schematic diagram of the beep transmitter. The components are placed on a printed circuit board as shown in Fig. 3.36.

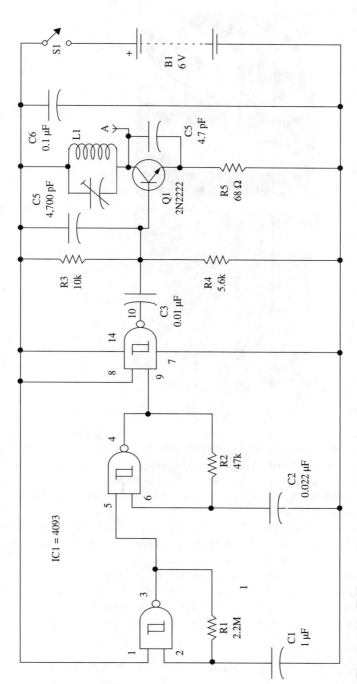

Figure 3.35   FM beep emitter.

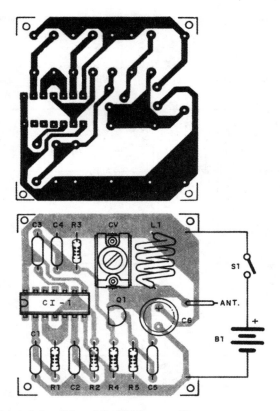

Figure 3.36   Printed circuit board for FM beep emitter.

For operation in the FM range, the coil is formed by four turns of AWG 18 to 22 enameled wire or plastic-covered common rigid wire wound around a pencil as a reference. For the VHF band between 50 and 80 MHz, the coil is formed by five or six turns. For the band between 108 to 150 MHz, the coil is formed by one or two turns of the same wire.

Any trimmer capacitor with a capacitance range between 2–20 and 4–40 pF can be used for this project. The antenna is a piece of rigid plastic-covered wire 10 to 40 in. long or a common telescoping antenna.

The circuit can be powered from four AA, C, or D cells in the basic version, but if the circuit is changed to run on 12 volts, a Nicad or automotive battery must be used. With a 6 V supply current drain is about 20 mA, but this current increases to up to 200 mA when the circuit is powered from a 12 V supply.

Adjustment and Use

Tune an FM receiver to a free point of the FM band and place it near the transmitter (3 to 4 ft). Power up the transmitter and adjust CV to tune the signal.

Be sure that the tuned signal is the fundamental one. If you move the receiver a good distance from the transmitter and the signal doesn't disappear, then it is the fundamental.

Parts List: Project 21

Semiconductors

IC1   4093 CMOS integrated circuit

Q1   2N2222, BF494, or 2N2218 high-frequency NPN silicon transistor

Resistors (1/8 W, 5%)

R1   2,200,000 Ω—red, red, green

R2   47,000 Ω—yellow, violet, orange

R3   10,000 Ω—brown, black, orange

R4   5,600 Ω—green, blue, red

R5   68 Ω—blue, gray, black

Capacitors

C1   1 μF metal film or electrolytic

C2   0.022 μF ceramic or metal film

C3   0.01 μF ceramic

C4   4,700 pF ceramic

C5   4.7 or 5.6 pF ceramic (see text)

C6   47 μF/12 WVDC, electrolytic

CV   trimmer capacitor (see text)

Additional Parts and Materials

B1   6 V, four AA cells (see text)

S1   SPST toggle or slide switch

L1   coil

Printed circuit board, battery holder, plastic box, wires, antenna, solder, etc.

4

Medium-Wave and Shortwave Transmitters

This chapter is dedicated to experimental low-power and medium-power, medium-wave (MW) and shortwave (SW) transmitters operating in the range between 550 kHz and 30 MHz. These transmitters can be used for short-range communications, using devices such as wireless microphones, or as wireless audio links to send sounds from one point to another in your home. The circuits can also be used in school or for scientific experiments. They are useful for learning a great deal about the principles of radio transmission, and they use easy-to-find components as bipolar transistors, FETs, and vacuum tubes.

Project 22: Small AM Transmitter

This one-transistor transmitter can be used as a wireless microphone for sending signals to any receiver placed at distances up to 15 ft.

Features

- Power supply: 6 to 9 Vdc
- Frequency range: 530 to 1,600 kHz
- Range: 6 to 15 ft

Small MW transmitters are easy to build but have some limitations in terms of range, as low-power signals cannot carry over large distances. Therefore, a small MW transmitter hasn't the same range as an equivalent FM transmitter using the same components and generating the same output power.

This experimental MW transmitter can be used to send signals to a nearby receiver, and it has some interesting applications as suggested below:

- As a wireless microphone, sending the signals to a nearby tape recorder or amplifier near it
- To transfer signals from one room to another, passing through the walls

- To demonstrate how a medium-wave station works for science fairs or in a school's technology education program

The circuit is powered from four AA cells or a 9 V battery, and the sound comes from a high-impedance microphone or other source.

How the Circuit Works

The circuit is a simple one-stage Hartley oscillator running in a range between 530 and 1,600 kHz. The frequency is determined by L1 and CV. CV is a common MW variable capacitor, and it is used to find a free point in the MW band. The feedback that keeps the circuit running is derived from tap 1 in the coil and capacitor C2.

Assembly

A schematic diagram of the circuit is shown in Fig. 4.1. As the circuit is experimental and intended for beginners and students, the components can be mounted on a terminal strip as shown in Fig. 4.2.

The audio input connection is generally made by a jack and plug. However, if the circuit is used as a wireless microphone, these components are not necessary, and the microphone can be connected directly to the circuit.

The capacitors are ceramic types. Any variable capacitor with capacitances ranging between 120 and 360 pF can be used. The antenna is a piece of rigid plastic-covered wire 10 to 40 in. long or a telescoping antenna.

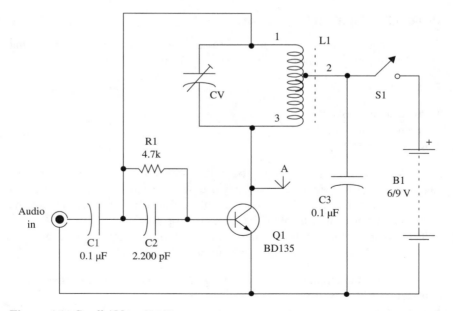

Figure 4.1 Small AM transmitter.

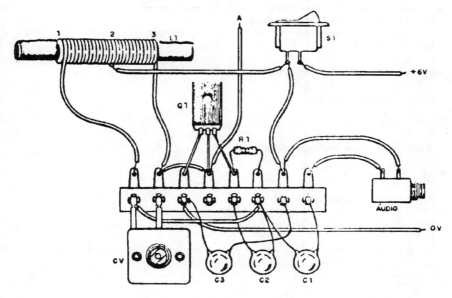

Figure 4.2   Mounting the circuit using a small terminal strip as the chassis.

The coil is formed by 40 + 40 turns of AWG 26 to 30 enameled wire over a ferrite core with diameter between 1/4 and 1/2 in. (0.8 to 1.5 cm) and from 4 to 10 in. (10 to 25 cm) in length. The circuit can be installed in a small plastic box.

Adjustments and Use

Place any MW receiver near the transmitter and tune it to a free point in the band. Power the transmitter up and adjust CV to find the signal. You must make this adjustment while speaking into the microphone.

When using the unit, keep the antenna in the vertical position, and don't shake the device. For an experimental broadcast station, the antenna can be a piece of common plastic covered wire 3 to 10 ft long. A ground connection made at the negative pole of the power supply will increase the circuit range.

Parts List: Project 22

Semiconductors

Q1   BD135 or equivalent medium-power NPN silicon transistor

Resistors (1/8 W, 5%)

R1   4,700 Ω—yellow, violet, red

Parts List: Project 22 (continued)

Capacitors

C1   0.1 μF ceramic or metal film

C2   2,200 pF ceramic

C3   0.1 μF ceramic

Additional Parts and Materials

L1   coil (see text)

S1   SPST toggle or slide switch

B1   6 to 9 V, AA cells or battery

Printed circuit board, battery holder or clip, antenna, plastic box, input jack and plug or microphone, knob for the variable capacitor, wires, etc.

Project 23: Medium-Wave Repeater

This transmitter can be used to send the sound from an amplifier or other audio equipment to a receiver placed on the other side of a wall to drive a remote loud-speaker in a sound system.

Features

- Power supply voltage: 6 Vdc
- Frequency range: 530 to 1,600 kHz
- Used configuration: Hartley oscillator
- Number of transistors: 1

Using this low-power MW repeater, you can send the output of a sound system to a loudspeaker placed on the other side of a room, or in another room, without the need of wires. The circuit is a low-power AM transmitter driven directly from the audio output of any sound equipment. As the signals produced in the MW range cannot carry a great distance, the circuit is useful only for short-distance links.

Figure 4.3 shows how this circuit can be used in the suggested application. You can also use the circuit to send the TV audio channel to a portable MW receiver using an earphone. Using this, it is possible to watch your favorite programs late at night without disturbing neighbors or other people in your home.

Another use for this circuit is as an experimental MW transmitter for the home. You can send the program from a CD player or a tape recorder to MW receivers in other rooms. The circuit is very simple and uses only one transistor.

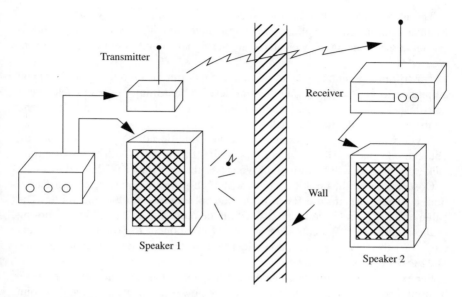

Figure 4.3   Using the medium-wave repeater.

How It Works

A transistor is wired as a Hartley oscillator running at a frequency between 530 and 1,600 kHz in the medium-wave band. The coil L1 and CV1 determine the frequency. The user must adjust CV1 to find a free point in the MW band.

The output power depends on the supply voltage and also on the antenna. There are two possible antenna configurations. One uses the coil's ferrite core as the antenna. In this case, the signal will be sent to distances of only a few feet as in a through-wall link.

The other possibility is to send the signals through the ac power line. In this case, the ac power wiring of a house acts as an antenna, allowing the transmitter to cover large areas. You will be able to pick up the signal at various places in your home by placing the receiver near any power outlet, switch, or wire used for the ac power line.

You can power the unit with four AA cells or a small power supply from the ac power line. But if you increase R1 to 10,000 Ω, you can power the circuit from 9 or 12 V supplies.

The value of resistor R2 depends on the transistor gain. We recommend that the reader experiment, changing this component in the range between 2,200 and 10,000 Ω, to find the best performance.

Any NPN medium- or high-power transistor that is suitable for use as an oscillator in the MW range (e.g., the BD135, TIP31, and 2N3055 components) can be used in this circuit.

The low-frequency modulation comes from the output of any audio system. You can pick up this signal from the earphone or monitor output. Using a transformer at the input signal, we can isolate the signal source from the circuit for increased security.

The trimmer potentiometer wired to the transformer is used to adjust the signal amplitude to get the correct modulation level. This trimmer potentiometer must be adjusted to get the best modulation without any distortion, which will depend on the volume of the input signal.

If you want to use the circuit with devices that do not have an audio output, you can add one as shown in Fig. 4.4. Using this circuit (with a closed-circuit jack), the speaker will be turned off whenever the transmitter is plugged in.

Observe that, if the circuit operates with high-power signal source for modulation, it is necessary to use Rx to limit the applied power to the transmitter. The value of R1 is found experimentally within the range from 22 to 220 Ω according to the input power. We can suggest a 47 Ω component to power signals in the range of 1 to 10 W. For a common television set, this resistor will be in the range of 22 to 39 Ω.

For experimental purposes, a small antenna wired as shown in the diagram can be used. If the circuit uses the wires of the ac power line as its antenna, this point should be connected to the ground pole of the line.

Assembly

The complete schematic diagram of the MW repeater is shown in Fig. 4.5. The project can be mounted on a printed circuit board as shown in Fig. 4.6.

The coil is formed by 40 + 40 turns of AWG 26 to 30 enameled wire in a small ferrite core, 4 to 10 inches long. The diameter can be in the range of 1/4 to 1/2 in.

The variable capacitor is a 120 to 360 pF type. Any variable capacitor found in a nonfunctioning old MW radio can be used for this circuit. But the reader must avoid using components extracted from FM radios, as they have lower capacitance ranges. If a low-capacitance variable capacitor is used, the frequency band covered will be narrow, and it will be difficult to tune the signal.

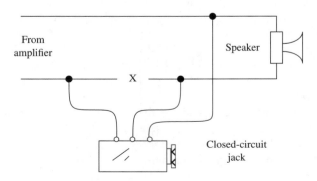

Figure 4.4 Adding an output to a sound system.

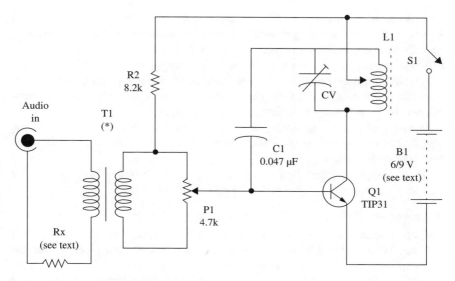

Figure 4.5   Medium-wave repeater.

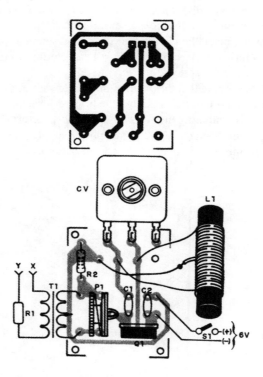

Figure 4.6   Printed circuit board for Project 23.

This circuit can use several types of transformers. The only requirement is that it must have a high-impedance winding and a low-impedance winding to be suitable for the task. The low-impedance winding is wired to R1.

In the prototype, for instance, we used a 117 Vac primary and 6 V × 250 mA secondary winding transformer. The high-voltage primary winding is the high-impedance winding. The reader can conduct experiments with available transformers to identify the unit with the best performance.

If the signal power is in stereo, the reader must use a special plug to pick up the signals that will be carried to the transmitter. Such a plug, with the necessary connections, is shown in Fig. 4.7.

The circuit can be installed in a plastic or wooden box. Don't use a metallic box, as it can block the signal produced by the circuit.

Adjustments and Use

Plug the circuit into an audio source that has been adjusted to a medium output volume. Tune a MW receiver to a free point in the band and place it near the transmitter (3 to 4 ft away). Turn the transmitter on and adjust CV until the signal is tuned in. Then, using P1, adjust the modulation to obtain a pure reproduction in the receiver.

Move the receiver some distance from the transmitter to see how far the signal carries. It will be tunable to a distance of several feet.

To use the project, place the circuit near the sound source or the wall through which the signals must be transmitted. Adjust P1 according to the power source and tune the receiver to the signals.

It is important to observe that the position of the transmitter's coil in relation to the receiver coil is important for obtaining the best performance. Try several positions for the receiver, aligning the coils to get the best performance.

You can run the unit from a 6 V × 250 mA power supply that draws its current from the ac power line. This will avoid the cost of batteries.

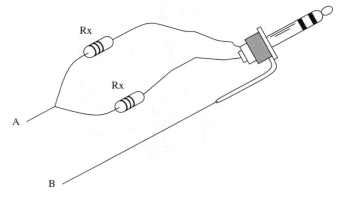

Figure 4.7   Using a stereo source to drive the transmitter.

Parts List: Project 23

Semiconductors

Q1   TIP31 or equivalent power NPN silicon transistor

Resistors (1/8 W, 5%)

R1   (see text) 1 W

R2   8,200 Ω—gray, red, red

P1   4,700 Ω trimmer potentiometer

Capacitors:

C1   0.047 µF ceramic

C2   0.1 µF ceramic

Additional Parts and Materials:

T1   transformer: 117 Vac primary winding; 6 V to 12 V × 250 to 500 mA secondary winding (see text)

L1   coil (see text)

CV   120 to 360 pF variable capacitor (see text)

B1   6 V, four AA, C, or D cells or power supply

Printed circuit board, plastic box, battery holder, ferrite core for the coil, wires, jack, etc.

Project 24: MW Radio Link

This circuit can be used to send signals from a sound source (amplifiers, CD-players, tape recorders) to any MW receiver transferring them to amplifiers or other sound systems without the need of wires. The circuit can also be used as a small MW experimental radio station.

Features

- Power supply: 117 Vac
- Frequency range: 530 to 1,600 kHz
- Output RF power: 1 W (typical)
- Modulation power recommended: 50 mW
- Range: up to 120 ft

A signal repeater with a respectable output power, using the MW radio band, can be used in several interesting applications. The following are some suggestions for the reader:

- When driven by the output of any audio system, the sounds can be transmitted to a receiver that is plugged into a remote audio amplifier. A wireless sound distribution system can be created in this manner.
- Using the output of any small audio amplifier with a hidden microphone, the transmitter can send signals to a receiver in an adjacent room. In this way, conversations can be overheard at a distance.
- Used with a tape recorder or mixer, the circuit can function as an experimental MW radio station for a school, club, or other organization. It can be used in science fairs to demonstrate how a radio station works.

The signal range depends on several factors. One of them is the presence of obstacles that can inhibit signal propagation. Walls and large metallic structures can block the signals. Another factor is the receiver's sensitivity.

As in the MW band, the signals cannot carry over large distances. The average maximum distance for this project is 120 ft. This range can be increased if the ac power wiring is used as the antenna.

The circuit is very easy to build and has no critical adjustments. This makes the project ideal for beginners and students.

How It Works

A single transistor is used in a Hartley oscillator. The frequency is determined by CV1 and L1. CV1 must be adjusted to find a free point in the MW band between 530 and 1,600 kHz.

The feedback that keeps the oscillator running is provided by C1. R2 biases the transistor through P1. P1 also adjusts the amplitude of the modulating signal.

If the amplitude is very high, the signal peaks are chopped, causing distortion. Therefore, the ideal adjustment point for this component depends on the power of the audio signal source.

The transformer is important for isolating the audio source circuit from the transmitter circuit, increasing security, and at the same time matching their impedances to obtain the best performance.

As the current drain is very high, a power supply that uses the ac power line is included with this project. This power supply uses a transformer with a primary winding rated to the 117 Vac power line (or 220/240 Vac, if this is the voltage where you live) and a secondary winding rated to 12 V CT, 1 A.

Capacitors C2 and C3 are used as filters. It is important keep all the wires short when building this circuit to avoid hums and instability. Special care must be taken with the length of the wire that is connected to CT in the coil.

Assembly

The circuit of the radio link is shown in Fig. 4.8. As this is an experimental project, we suggest that readers who are less experienced with printed circuit board etching should use a terminal strip as the chassis. Many variations of common terminal strips can be used. The components are placed on the terminal strip as shown in Fig. 4.9.

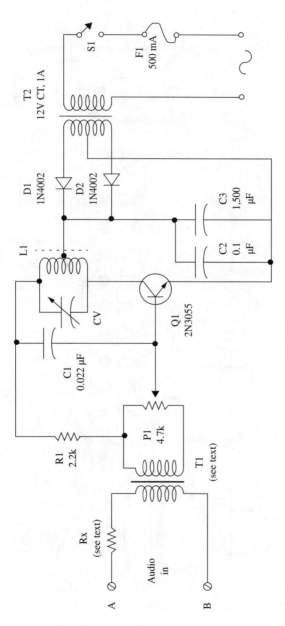

Figure 4.8   MW radio link.

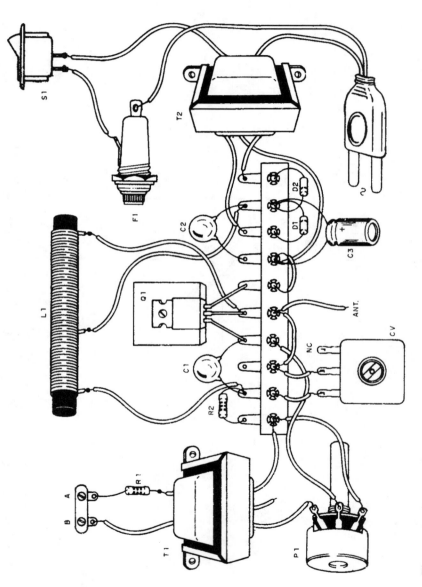

Figure 4.9 A terminal strip is used as the chassis for this project.

The coil is formed by 50 + 50 turns of AWG 26 to 30 enameled wire on a ferrite core 4 to 10 in. long. The diameter range is between 1/4 and 1/2 in. A suitable ferrite core can be found in old transistor radios.

For CV1, any variable capacitor with a capacitance range between 120 and 360 pF can be used. This component can also be found in old nonfunctioning transistor and tube radios. C1 and C2 must be ceramic capacitors, and C3 is an electrolytic rated to 25 WVDC or more.

For P1, any common linear or log potentiometer can be used. This component must be mounted appropriately so that the reader can use it to adjust the modulation.

The transistor must be mounted on a heat sink. Notice that if you use a 2N3055 or any equivalent in a TO-2 case, the terminal placement will be different from that shown, and the type of heat sink will also be different. The reader should be alerted to the fact that a 2N3055 cannot oscillate at frequencies higher than 1 MHz, which will be important for some applications. The circuit must be adjusted, when using this transistor, to a frequency between 530 and 650 kHz for best performance. The reader can use equivalents to the specified diodes such as the 1N4004 or 1N4007.

Many types of transformers can be used for this project. Any transformer with a high-impedance winding and a low-impedance winding (connected to R1) can be used. For the prototype, we used a transformer with a 117 Vac winding (wired to P1) and a 6 V × 250 mA secondary winding (plugged to the audio input).

R1 depends on the signal source. The following values are recommended:

Audio source power	R1
0 to 5 W	22 Ω × 1/2 W
5 to 20 W	47 Ω × 1 W
20 to 50 W	100 Ω × 1 W
50+ W	150 Ω × 1 W

As many audio amplifiers are rated to PMPO (peak power) the correct value for R1 must be determined experimentally. A 20 W (PMPO) amplifier can in fact provide only 5 W (RMS) to a load.

Adjustments and Use

For experimental applications, the antenna is formed by a long piece of plastic-covered wire (6 to 20 ft) placed in a horizontal position as shown in Fig. 4.10. You can also use the ac power line as an antenna. Connect the antenna terminal to the ac power line using a 0.01 μF × 400 V metal film capacitor. Try the two poles to see which one provides better performance.

When using the ac power line as the antenna, or even using a wire antenna if the receiver is placed near any wire connected to the ac power line (i.e., near outlets or light switches), the signal can be picked up easily.

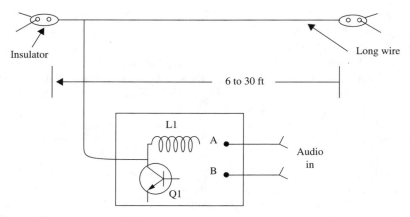

Figure 4.10 Adding an antenna.

Adjustment is easy: place any MW receiver near the transmitter and tune it to a free point in the MW band. Plug an audio source into the input of the transmitter and adjust CV to tune the signal. Adjust P1 and the audio source volume control to get the best sound without distortion. The transistor will be warm in operation. This is normal.

Parts List: Project 24

Semiconductors

Q1   TIP31, TIP3055, or 2N3055 high-power NPN silicon audio transistor

D1, D2   1N4002 or equivalent silicon rectifier diodes

Resistors (1/8 W, 5%)

R1   (see text)

R2   2,200 Ω—red, red, red

P1   10,000 Ω—potentiometer

Capacitors

C1   0.022 µF ceramic

C2   0.1 µF ceramic

C3   1,500 µF/25 WVDC electrolytic

Additional Parts and Materials

T1   High-to-low-impedance transformer (see text)

T2   Transformer: 117 Vac primary winding; 12 V CT secondary winding rated to currents of 1 A or more

---

Parts List: Project 24 (continued)

S1   SPST toggle or slide switch

F1   1 A fuse and holder

Printed circuit board, power cord, plastic box, heat sink for the transistor, wire, antenna, ferrite core, solder, etc.

Project 25: Two-Transistor MW Transmitter

This small MW transmitter uses two audio transistors to create a strong signal between 530 and 1,600 kHz. The circuit can be used as an experimental MW radio station.

Features

- Power supply voltage: 6 to 9 Vdc
- Range: 15 to 100 ft
- Frequency range: 530 to 1,600 kHz
- Oscillator: Hartley

Two transistorized stages are used in this transmitter to increase the output power. But because, in the MW range, it requires a great deal of power to send the signal over long distances, this project does not provide the same performance as equivalent ones operating in the SW or FM range. It therefore can be used as an experimental radio station or link to send audio signals to distances only up to about 100 ft.

Several uses are suggested for a circuit like this. The applications are the same as described for similar projects as follows:

- It can be used as a radio link to send signals from a tape recorder, TV, or audio system to a remote receiver or loudspeaker.
- It may be useful as an experimental radio station for sending programs to radios placed around the school or home.
- It can be employed as a secret listening device when combined with an amplifier and a microphone.

The circuit can be powered from 6 to 9 V power supplies. The supply can use single cells or a battery. If you use a supply that is fed from the ac power line, a 6 to 9 V × 500 mA rated type is recommended.

How It Works

A high-frequency signal in the range between 530 and 1,600 kHz is produced by a Hartley oscillator using a medium-power silicon transistor. The signal frequency is determined by L1 and CV1. CV1 must be adjusted to a free point in the MW band.

This circuit performance is improved by a high-frequency amplifier stage. This stage uses another medium-power NPN silicon transistor and can increase the signal output to as much as 1 W.

Modulation is achieved by the aid of a transformer, as in the other MW amplitude modulated (AM) circuits described in this book. The transformer must match the audio system output impedance to the oscillator stage input impedance.

A potentiometer is used to adjust the correct signal level to be applied to the transistor base. This adjustment is necessary to find the best modulation without any distortion.

The second stage is a class C stage, where there is no resistor to bias the transistor. The transistor is biased by the charging and discharging process of C2. This means that the transistor conducts only during one of the signal half-cycles.

As the load for this stage, we use a high-frequency choke. This means that the signal can be picked up from the second-stage collector.

The standby current for the second stage is given by R1 and R2. For R1, the reader can conduct experiments with resistors between 2,200 and 10,000 Ω to find the best performance.

We also include a second modulation input to the second stage. Low-impedance signals using a transformer can be applied at this point, modulating the signals. Using a transformer (4 or 8 Ω secondary winding), the output of a small amplifier in the power range of 200 mW to 2 W can be connected to this point.

The reader is free to make many modifications in this circuit, such as

- Alter R1 and C1 for better performance.
- Change R2 for better modulation.
- Alter L1 according to the frequency.

Assembly

A schematic diagram of the two-stage MW transmitter is shown in Fig. 4.11. As the circuit is not critical and is designed for students and beginners, the components can be mounted on a terminal strip as shown in Fig. 4.12. Of course, a reader who knows how to etch a printed circuit board may prefer that option, which allows the unit to be installed in a smaller box.

The coil is formed by 50 + 50 turns of AWG 26 to 30 enameled wire over a ferrite core 4 to 10 in. long. The diameter can be in the 1/4 to 1/2 in. range. If you have trouble finding AWG enameled wire, use common plastic-covered AWG 22 wire to wind the coil.

For CV, you can use any variable capacitor with capacitances in the range between 120 and 360 pF. This capacitor can be found in old transistorized MW radios. Avoid variable capacitors found in FM radios, as they are low-capacitance types.

L2 is a high-frequency coil with an inductance between 47 and 470 μH. You can create a home-made version by winding 200 to 300 turns of AWG 30 to 32 enameled wire on a 100 k$\Omega \times$ 1 W resistor.

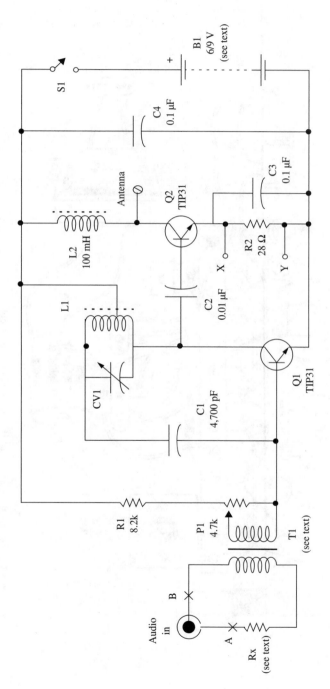

Figure 4.11   Two-transistor MW transmitter.

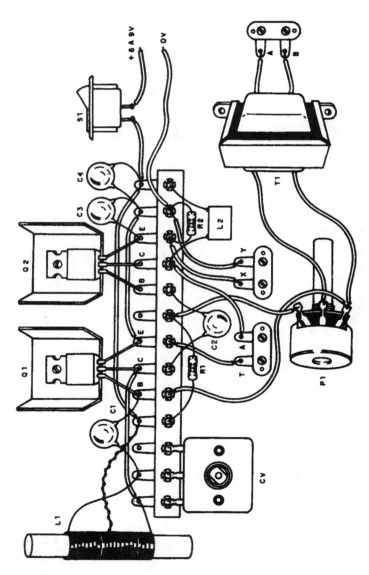

Figure 4.12   Terminal strip used as the chassis in this assembly.

Any transformer with a high-impedance primary winding and a low-imped-ance secondary winding can be used for modulation. In the prototype, we used a 117 Vac primary winding and 6 to 12 V secondary winding with current ratings in the range between 250 and 500 mA. The low-voltage secondary is plugged into the audio input. You can also use a small audio output transistor as found in old transistor radios for this task.

The transistor must be mounted on heat sinks. The heat sinks consist of small pieces of metal bent to form a "U" and affixed to the transistors.

A power supply for this circuit is shown in Fig. 4.13. The transformer has a 117 Vac primary winding and a secondary winding rated to 6 V CT with currents of about 250 mA or more.

The electrolytic capacitor must be rated to 12 V or more, and you can use equivalents to the diodes. A terminal strip with screws is used to connect the an-tenna and the ground wire. If you intend to use cells, C or D types are recom-mended, as the current drain is high.

Adjustments and Use

Figure 4.14 shows how the circuit can be used with a tape recorder to send music to remote receivers. If the reader intends to use the transmitter with an electret microphone, a small amplifier must be added as shown in Fig. 4.15. The amplifier should generate 100 mW to 1 W.

The antenna is formed using a piece of plastic-covered wire 6 to 30 ft long and wired as shown in Fig. 4.16. The ground connection is very important for increas-ing performance. This connection can be made to the neutral pole in the ac power line as shown.

To adjust the circuit, place the MW receiver near the transmitter and tune it to a free point in the band. Power up the transmitter and adjust CV1 to pick up the

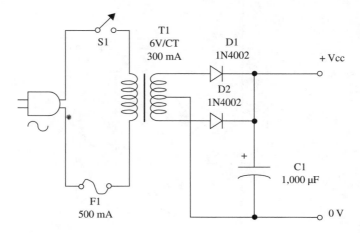

Figure 4.13 A suitable power supply for this transmitter.

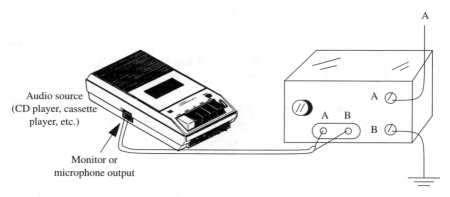

Figure 4.14   Sending signals from a tape recorder.

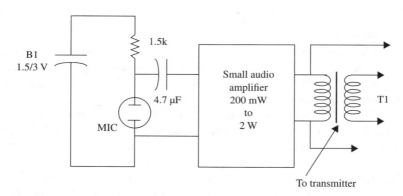

Figure 4.15   Using an electret microphone.

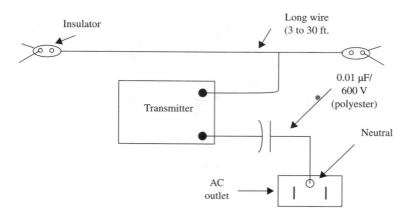

Figure 4.16   Wiring an antenna and making a ground connection.

signal, and P1 to get the best sound reproduction without distortion. A tape recorder, CD player, or other audio source can be used for the adjustments.

Parts List: Project 25

Semiconductors

Q1, Q2   TIP31 medium power NPN silicon transistor, any suffix

Resistors (1/8 W, 5%)

R1   8,200 Ω—gray, red, red

R2   22 Ω—red, red, black

P1   4,700 Ω—potentiometer

Capacitors

C1   2,200 μF ceramic

C2   0.01 μF ceramic

C3, C4   0.1 μF ceramic

CV   variable capacitor (see text)

Additional Parts and Materials

T1   transformer (see text)

L1, L2   coils (see text)

B1   6 to 9 V, cells or power supply (see text)

S1   SPST toggle or slide switch

Terminal strip, plastic box, power cord (if necessary), components for the power supply, battery holder, ferrite core, knobs for the variable capacitor and the potentiometer, wires, solder, etc.

Project 26: Shortwave Telegraphic Transmitter

This very low-power transmitter can be used to send telegraphic signals to a receiver placed at a distance up to 100 ft. This simple experimental circuit is useful for demonstrations of radio principles.

Features

- Power supply voltage: 3 to 9 Vdc
- Frequency range: 3 to 15 MHz
- Range: 100 ft
- Transmission mode: CW (continuous wave)

This small transmitter sends a signal in the shortwave band between 3 and 15 MHz, depending on the selected coil. The circuit can be used in emergency situations or for demonstrations. Depending on the antenna used and the receiver's sensitivity, the signals can be received at distances up to several hundred feet. As this is an experimental project, one of the suggested uses is for a science fair to demonstrate how wireless telegraphy operates.

The basic circuit uses a BF494 and is powered from a 3 to 6 V supply, but it can be altered to use a BD135 or 2N2218 and be powered from a 12 V supply. With these changes the power will be increased and also the range.

How It Works

The circuit is formed by a one-transistor oscillator where the frequency is determined by L1 and CV. The high-frequency transformer secondary (L2) provides the necessary feedback to keep the transistor in oscillation.

R1 is the component that biases the transistor, and it can be altered within a wide range of values. The reader can experiment with values in the range between 2,200 and 22,000 Ω.

The antenna is connected to the transistor's collector. For best performance, a good ground connection is important.

For experimental purposes, the antenna can be a piece of plastic-covered wire 10 to 40 ft long, and the ground connection can be made to any large, metallic object that is in electrical contact with the ground.

Assembly

Figure 4.17 shows a schematic diagram of this shortwave experimental transmitter. Because the circuit is very simple, the components can be mounted on a terminal strip as chassis. This assembly is shown in Fig. 4.18.

The coils are wound on a small ferrite core 3 to 6 in. long. The diameter can be in the range between 1/4 and 1/2 in.

L1 is formed with 30 turns of AWG 22 to 28 enameled wire, and L2 is formed with 10 turns of the same wire for operation in the range between 3 and 7 MHz. For operation between 7 and 15 MHz, L1 is formed with 10 to 12 turns and L2 with 4 to 6 turns of the same wire. Note that L2 is wired over L1. If the circuit doesn't oscillate when you power it on, invert one of the coils.

The power supply consists of four AA cells in the basic version, but the circuit can operate from 3 to 9 V supplies. Any variable capacitor with capacitances ranging between 120 and 360 pF can be used for this circuit.

For the basic version, you can use the BF494, but you can get more power with transistors such as the BD135 and TIP31. But when using these transistors, you must verify the terminal positions, as they are different. These transistors must be mounted on small heat sinks if the circuit is powered from 9 to 12 V supplies.

The morse key can be a homemade type as shown in Fig. 4.18. A metallic blade is bent to touch a metal contact (a screw, for instance), and both are affixed on a wooden base.

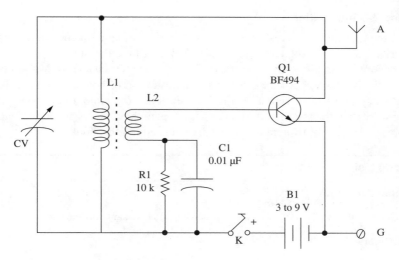

Figure 4.17   SW telegraphic transmitter.

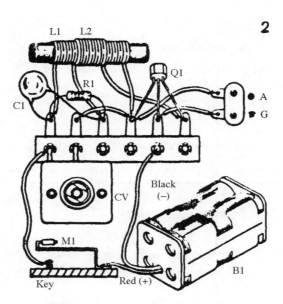

Figure 4.18   Circuit assembly using a terminal strip as chassis.

Another noncritical component is the ceramic capacitor C1. The reader can experiment with values between 2,200 pF and 0.022 µF for this component.

The antenna is a piece of plastic-covered wire 6 to 60 ft long. The ground connection is very important for increasing performance.

Using the Transmitter

Near the transmitter, place an SW receiver tuned to a free frequency in the range determined by the coil. Press the Morse key and tune the signal by adjusting CV1. For this circuit to have any practical use, the reader must know Morse code.

With this device, you can conduct a demonstration of how a traditional telegraphic system works as well as this wireless one. You simply have to explain how words are translated into Morse code. Explain how this is accomplished. A Morse code table is given in the Appendix at the end of this book.

Parts List: Project 26

Semiconductors

Q1   BF494, 2N2218, BD135 or equivalent NPN silicon transistor (see text)

Resistors

R1   10,000 Ω—brown, black, orange

Capacitors

C1   0.01 μF ceramic

CV   variable capacitor (see text)

Additional Parts and Materials

L1, L2   coils (see text)

K   Morse key (see text)

B1   3 to 9 V - AA cells, battery or power supply (see text)

Terminal strip, battery holder, ferrite core, antenna, wires, solder, etc.

Project 27: One-Transistor Shortwave Transmitter

This experimental shortwave transmitter can be modulated by audio sources sending the signals to distances up to several hundred feet.

Features

- Power supply voltage: 9 to 12 Vdc
- Output power: 500 mW to 2 W
- Frequency range: 3 to 10 MHz
- Number of transistors: 1
- Range: up to 300 ft (see text)

This project is intended for beginners and students who are looking for a circuit for conducting experiments in shortwave transmission. The circuit is simple and has no critical parts, so it can use a terminal strip as its chassis.

Because the shortwave range is less congested during daylight hours, this is the best time to locate a free point in the band for the operation of this device. Obviously, this transmitter requires you to have access to a shortwave receiver for it to be of any use.

The modulation can be created by any small audio amplifier. The reader can use the circuit with any audio source, including an electret microphone and other types of microphones and transducers.

It is also important to remind the reader that, if an external antenna is used, the signal can carry over long distances. This means that legal restrictions on its operation must be carefully observed. For experimental purposes, the reader must use only the recommended antenna.

How It Works

The transistor is wired as a Hartley oscillator running at a frequency determined by L1 and CV1. The circuit can be tuned by CV1 to a free point in the shortwave band between 3 and 10 MHz.

The feedback necessary to keep the oscillator running is given by C1. R1 wired in parallel with C1 is used to bias the transistor base.

To modulate this signal from an audio source, a transformer is used. One winding is used to control the power supply current, and the other to receive the audio signal.

Without any audio signal applied, the current to the circuit flows without any resistance. But, when a signal is applied, the resulting voltage controls the current flow, causing the high-frequency modulation.

Figure 4.19 shows how the amplitude of the high-frequency signal must change with the audio signal to achieve a correct modulation.

Notice that, in this circuit, to have a good modulation, it is necessary to use a high-power audio signal. The audio signal must have the same power as the signal produced by the transmitter. This means that amplifiers with powers in the range between 500 mW and 2 W must be used for the external modulation.

Another away to modulate this transmitter is shown in Fig. 4.20. A small power supply transformer is used to control the current flowing to the transmitter.

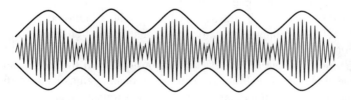

Figure 4.19 High-frequency signal modulated in amplitude.

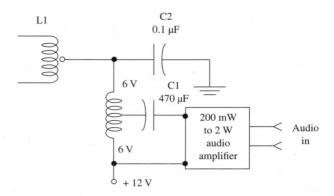

Figure 4.20   Another circuit for amplitude modulation.

Only the low-voltage winding is used. The applied signal acts on the current flowing and changes it, thereby altering the amplitude of the high-frequency signal produced by the Hartley oscillator. Given that the project is experimental, we invite the reader to test several transformers with the employed modulation amplifier to find the one that gives the best performance.

The signal produced by the high-frequency oscillator and subsequently amplitude modulated is picked up from the circuit by L2. This coil, via CV2, applies the high-frequency signal to an antenna and a ground connection. CV2 can be adjusted to match the circuit impedance, with the antenna impedance increasing the transmitter performance.

For good performance, you must use a power supply ranging from 9 to 12 V and current up to a level of 1 A. Good filtering is necessary to avoid hum.

Assembly

A schematic diagram of the shortwave transmitter is shown in Fig. 4.21. Because the project is experimental and not very critical, a terminal strip can be used as the chassis as shown in Fig. 4.22. Keep all the connections short to avoid instability, and take care not to touch any component terminals to any others.

L1 is formed by 15 + 15 to 20 + 20 turns of AWG 24 to 28 enameled wire on a ferrite core 4 to 10 in. long. The diameter can be in the range between 1/4 and 1/2 in. This ferrite core can be found in any old transistor radio. L2 is formed by 10 to 12 turns using the same wire over L1.

CV1 and CV2 are small, plastic variable capacitors in the range between 120 and 360 pF. These components can also be found in old transistor radios. But do not use FM radio parts, which have less capacitance and therefore will not work in the frequency range of interest for this project. With the lower capacitance range, the circuit will tune only a narrow band, making adjustments difficult.

Another method of tuning the circuit is to use trimmer capacitors with capacitance ranges between 50 and 120 pF. The frequency range will be narrow, but you

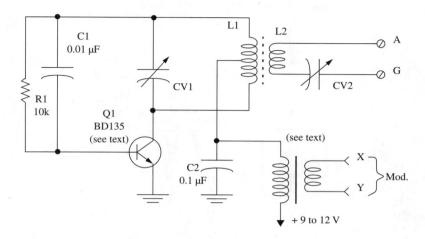

Figure 4.21   Shortwave transmitter.

should be able to find it somewhere in the shortwave receiver. If not, alter the number of turns of L1 until you can tune in the signal.

The transistor must be mounted on a heat sink. This can be a small metal plate bent to form an "L" or a "U" and affixed to the component with a screw and a nut.

Any transformer with a primary winding rated to voltages between 117 Vac and 240 Vac and a secondary winding rated to voltages between 5 and 12 V, with a current rating between 250 and 800 mA, can be used as the modulation circuit.

It is also possible to use small output transformers such as the ones found in transistor radios. The 4 or 8 Ω winding is wired to the output of the modulation amplifier, and the 200 to 500 Ω winding is wired to the transmitter.

Figure 4.23 shows a modulation circuit using an electret microphone. The gain is adjusted by P1. This modulator must be powered from two or four AA cells.

Adjustment and Use

Place any shortwave receiver near the transmitter and tune it to a free point in the shortwave band between 4 and 5 MHz. Use a small piece of plastic-covered wire (10 to 20 in. long) as the antenna for the transmitter.

Power on the transmitter and adjust CV1 to tune the signal. Apply a modulation signal and adjust the amplifier's volume to get the best sound reproduction without any distortion.

When using the circuit with an external antenna, also adjust CV2 to get the best performance. Figure 4.24 shows an external antenna suitable for this circuit.

Notice that the antenna is a half-wave dipole, so a ground connection is not used. The best performance for the transmitter is found when the antenna dimensions correspond to half the length of the transmitted signal.

The reader can experiment with alterations in the value of R1 to get the best performance. Values between 4,700 and 22,000 Ω can be tested.

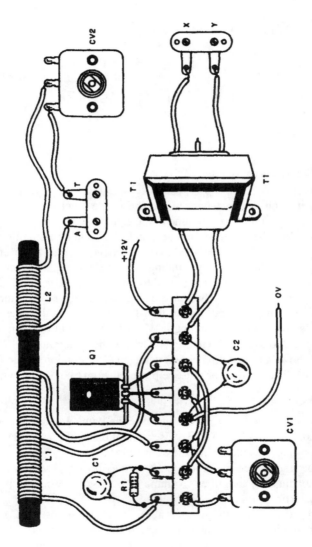

Figure 4.22 Experimental transmitter using a terminal strip as the chassis.

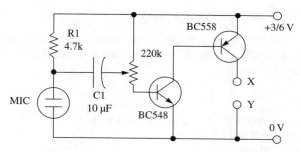

Figure 4.23   A simple modulator.

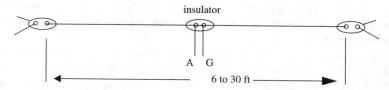

Figure 4.24   An external antenna.

Parts List: Project 27

Semiconductors

Q1   BD135, BD137, or BD139 medium-power NPN silicon transistor

Resistors (1/8 W, 5%)

R1   10,000 Ω—brown, black, orange

Capacitors

C1   0.01 µF ceramic

C2   0.1 µF ceramic

CV1, CV2   variables (see text)

Additional Parts and Materials

L1, L2   coils (see text)

T1   modulation transformer (see text)

Terminal strip, power supply, plastic box, knobs for the variable capacitors, audio source, wires, solder, etc.

Project 28: FET Shortwave Transmitter

This low-power shortwave transmitter is an experimental circuit using a field effect transistor (FET). The transmitter can be used as a wireless microphone or to test shortwave receivers.

Features

- Power supply: 6 to 9 Vdc (four AA cells or a battery)
- Frequency range: 7 to 15 MHz
- Output power: 20 mW (typical)

This very low-power shortwave transmitter can be used to send signals to a receiver placed 3 to 20 ft away. This means that it can be used as a wireless microphone or for demonstrating basic radio principles, such as in a science fair or in technology education.

However, as we described in Chapter 1, low power doesn't necessarily mean that signals cannot carry over long distances. Low-power transmitters can send signals to receivers a thousand miles away when using appropriate antennas and under special propagation conditions. Amateur radio operators can use these transmitters to test their skills in making contact with very distant radio installations. They often use transmitters in the range of 1 mW to 10 mW. It is a test of your technical skills to try to send the signals produced by this small transmitter over the greatest possible distance.

How It Works

This circuit uses a Hartley oscillator with an FET. The frequency is determined by L1 and CV. CV adjusts the frequency to a free point in the shortwave range between 7 and 15 MHz. R1 biases the transistor gate, and C1 applies the feedback signal to the transistor, which keeps the oscillator running.

For an experimental application, the signals are applied to a telescoping antenna 20 to 40 in. long. But, if the reader prefers, an external antenna can be connected to this point. In this case, it is important to establish a good ground connection to increase the signal propagation into space.

Figure 4.25 shows how a dipole or a long wire antenna can be used with this transmitter. The modulation is given by a high-impedance (crystal or ceramic) microphone placed as shown in the figure. Unfortunately, this circuit cannot use other types of microphones without modification. As the circuit is intended as a portable and experimental unit, this configuration is suitable for the task. Coil XRF1 is used as a load for the RF signals.

Assembly

The circuit of the FET shortwave transmitter is shown in Fig. 4.26. If you are a beginner or don't have the resources to etch a printed circuit board, you can use a terminal strip as shown in Fig. 4.27.

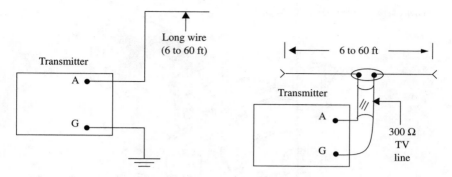

Figure 4.25   Using external antennas.

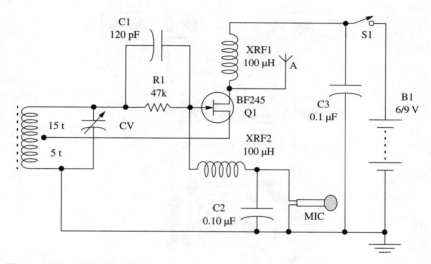

Figure 4.26   FET shortwave transmitter.

You can also mount the circuit on a printed circuit board. Figure 4.28 shows how the components are placed on it.

L1 is formed by 5 + 15 turns of AWG 26 to 30 enameled wire on a cardboard tube with a diameter of 1/4 in. Inside this tube, a small ferrite core must slide as shown in Fig. 4.29.

CV is a variable capacitor with capacitance in the range between 120 and 360 pF. Any type found in an old MW or SW transistor radio can be used.

It is also possible to experiment with a trimmer capacitor in the range between 2–20 and 3–30 pF, but in this case the frequency band covered by the transmitter will be as narrow as in the original version with a variable capacitor.

The original FET is a BF245, but equivalents such as the MPF102 can also be tested. The reader only has to pay attention to the terminal placement if an

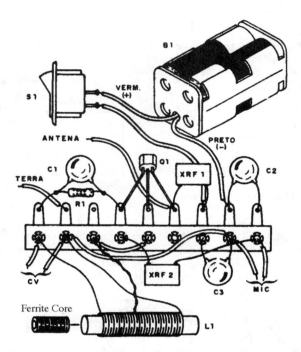

Figure 4.27 The circuit mounted on a terminal strip.

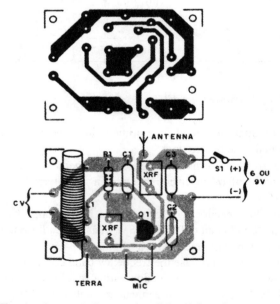

Figure 4.28 The circuit mounted on a printed circuit board.

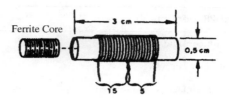

Figure 4.29   Coil details.

equivalent is used. For instance, the MPF102 does not have the same terminal configuration as the BF245.

XRF1 and XRF2 are RF chokes with values between 47 and 100 μH. But, as these components are not critical, a homemade component can be used. Wind 40 to 60 turns of AWG 30 to 32 enameled wire over a match or any plastic or cardboard tube with a diameter of 1 or 2 mm and a length of 2 to 3 cm.

The microphone must be a crystal or ceramic high-impedance type. Four AA cells or a 9 V battery can be used as the power supply.

All the components can be installed in a small plastic box. The basic dimensions to this box depend on the version and the power supply you use. Remember that a 9 V battery is smaller than four AA cells in a holder.

Adjustments and Operation

Tune a SW receiver to a free point in the band between 7 and 15 MHz. The receiver must be placed at a distance of 1 to 3 ft from the transmitter.

Power on the transmitter and adjust CV to tune the signal. At the same time, speak into the microphone to have a modulated signal to tune.

If your version uses an external antenna, connect it to the circuit and make new adjustments. The presence of an antenna can alter the frequency, requiring readjustment.

Be careful not to tune to a harmonic signal. The fundamental is the strongest one, and it can be tuned at the greatest distance from the transmitter. When using this as a portable unit, avoid shaking the antenna or touching components.

Parts List: Project 28

Semiconductors

Q1	BF245 or equivalent JFET

Resistor (1/8 W, 5%)

R1	47,000 Ω—yellow, violet, orange

Capacitors

C1	120 pF ceramic

Parts List: Project 28 (continued)

C2	0.01 µF ceramic
C3	0.1 µF ceramic
CV	variable capacitor (see text)

Additional Parts and Materials

MIC	Crystal or ceramic microphone (see text)
XRF1, XRF2	47 to 100 µF choke
B1	6 V or 9 V, four AA cells or 9 V battery
S1	SPST toggle or slide switch

Printed circuit board or terminal strip, battery clip or battery holder, antenna, plastic box, wires, solder, etc.

Project 29: Small, Two-Stage MW/SW Transmitter

This circuit can be used as a wireless microphone, for experimental purposes, or to listen to conversations through a wall.

Features

- Power supply voltage: 6 V to 12 V
- Frequency range: MW 530 to 1,600 kHz
            SW1 3 to 7 MHz
            SW2 7 to 15 MHz
- Range: up to 150 ft (depending on frequency)

This transmitter is powered from common AA cells and can send signals both in the MW range and the SW range, depending on the coil. Also depending on the coil (and the frequency range), the signals can be sent to receivers placed at distances up to 150 ft.

The circuit can be used as a wireless microphone, a mobile transmitter for short-range communications, and for demonstrations and experiments in technology education. A small, home-based MW radio station can also be made with this circuit.

How It Works

The circuit is formed by two stages: a modulation amplifier stage and a high-frequency oscillator. The modulation stage picks up the audio signals from an electret microphone. One transistor in the common emitter configuration is used in this stage. Resistor R1 biases the electret microphone, and R2 biases the transistor's base. Depending on the transistor gain, you must experiment with the value of resistor R2 in the range between 470,000 and 2,200,000 Ω to obtain the best performance without distortion.

The signals amplified by this stage are applied to the next stage by capacitor C2 and potentiometer P1. P1 can adjust the modulation level to find best performance.

The high-frequency stage is a Hartley oscillator, and its frequency is determined by L1 and adjusted by CV. Three options for the coil are suggested to the reader according to the choice of frequency range. These options are as follows:

Frequency range	Coil
550 to 1,600 kHz	40 + 40 turns
3 to 7 MHz	25 + 25 turns
7 to 15 MHz	18 + 18 turns

In many cases, you will need to add or reduce the number of turns in the coils to find the correct range.

All of the coils are formed by AWG 26 to 28 enameled wire on a ferrite core 4 to 10 in. long. The diameter is in the range between 1/4 and 1/2 in.

CV is any variable capacitor with capacitances in the range of 120 to 360 pF. Any variable capacitor found in an old MW or SW transistor radio can be used. Alternatively, a trimmer capacitor can be used for this project.

The signal can be conducted to the telescoping antenna from the transistor collector or, if you intend to use an external antenna, from a second coil wound over L1. This coil can be formed with 5 to 10 turns of AWG 28 enameled wire.

Assembly

A schematic diagram of the MW/SW transmitter is shown in Fig. 4.30. The components are mounted on a printed circuit board as shown in Fig. 4.31.

Details about the coil are given in the previous section ("How It Works"). The printed circuit board is intended for the use of a trimmer capacitor (between 2–20 and 8–80 pF), but the layout can be altered to accept a variable capacitor.

All capacitors with values below 1 µF are ceramic. For values above that level, the components are electrolytic capacitors rated to 12 WVDC or more. The reader must take care when mounting the electrolytic capacitors, as they are polarized components. Their proper positions must be observed.

Transistor Q2 must be installed on a small heat sink. This heat sink can be made from a 1 × 2 in. metal plate bent to forma a "U" and affixed to the transistor by a screw. Equivalents for Q2 are the BD137, BD139, or any other medium-frequency transistor with a frequency transition near 50 MHz or higher.

The antenna can be a piece of plastic-covered wire 10 to 40 in. long or an external antenna as described for the other shortwave transmitters in this book. A telescoping antenna can be installed if the circuit will be used for short-range communications or as an experimental radio station.

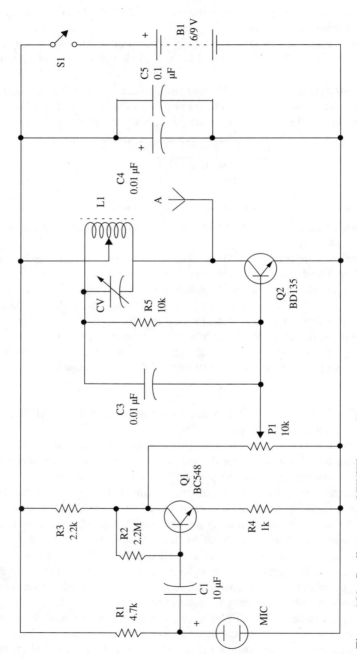

Figure 4.30 Small two-stage MW/SW transmitter.

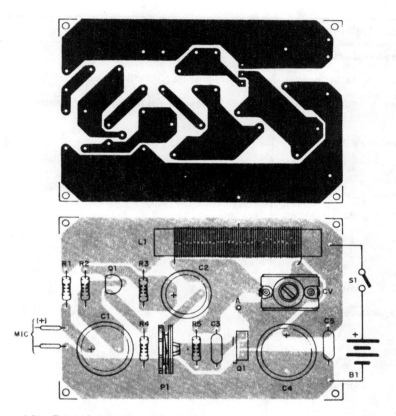

Figure 4.31   Printed circuit board for Project 29.

The circuit can be powered from 6 to 12 V supplies. If a 12 V supply is used, increase R1 to 10,000 Ω. For a 6 V supply, you can use AA, C, or D cells, but for 9 or 12 V you should use a high-power battery or a power supply that draws from the ac power line.

Adjustments and Use

Near the transmitter, place a receiver that will be tuned to a frequency that is determined by the specific coil used. Turn the transmitter's power supply on and adjust CV to tune in the strongest signal.

Then, speaking into the microphone, adjust P1 to get the best modulation. If the sound tends to distort, change the value of R2 as explained before.

If you intend to use an external audio source to modulate the circuit instead of the electret microphone, the signal can be applied to C1. R1 must be removed from the circuit in this case. Improved performance can also be achieved if you experiment with changing the value R5 within the range between 4,700 and 22,000 Ω.

Parts List: Project 29

Semiconductors

Q1   BC548 or equivalent general-purpose NPN silicon transistor

Q2   BD135 or equivalent medium-power audio NPN silicon transistor (see text)

Resistors (1/8 W, 5%)

R1   4,700 Ω—yellow, violet, red

R2   2,200,000 Ω—red, red, green

R3   22,000 Ω—red, red, orange

R4   1,000 Ω—brown, red, red

R5   10,000 Ω—brown, black, orange

P1   10,000 Ω—trimmer potentiometer

Capacitors

C1, C1   10 µF/12 WVDC electrolytic

C3   0.01 µF ceramic

C4   10 µF to 22 µF/12 WVDC electrolytic

C5   0.1 µF ceramic

CV   trimmer or variable capacitor (see text)

Additional Parts and Materials

MIC   electret microphone - two terminals

A   antenna (see text)

S1   SPST toggle or slide switch

B1   6 V to 12 V power supply, AA, C, or D cells, battery, etc. (see text)

Printed circuit board, plastic box, battery holder (if necessary), wires, ferrite core, solder, etc.

Project 30: High-Power AM Tube Transmitter

This high-power transmitter can be used as a small radio station for clubs, schools, and summer camps, sending a strong signal to receivers placed at distances up to half a mile, depending on the antenna.

Features

- Power supply voltage: 117 Vac
- Frequency range: 530 to 1,600 kHz
- Distance range: up to half a mile
- Number of tubes: 2

A very popular old tube receiver used in the 1960s and 1970s employed the 35W4, 12BE6, 12BA6, 6AV6, and 50C5 series of vacuum tubes. Those old receivers, tuning the MW and SW band, used the tubes with filaments wired in series as shown in Fig. 4.32. With this configuration, the receiver could be powered directly from the ac power line without the need of transformers, which were high-cost components at the time (and remain so today).

The filament voltage sum was 121 V—a little bit more than the power supply line voltage. So, with the use of a small resistor in series to add a voltage drop of a few more volts, the tubes could be heated directly from the ac power line.

In many such receivers, the resistor was inside the power cord (consisting of a resistive wire) so that when the receiver operated, the power cord was heated. Because of this, these receivers were often referred to as *heated tail* radios.

Today, it is possible to find many of these old receivers abandoned, with the tubes still in good shape. If the reader can find these tubes (and the other components as well), an interesting MW transmitter can be built. Using two of the tubes found in these receivers, we can build an "old-fashioned" transmitter that can send signals to MW receivers placed at distances of up to half a mile.

The 50C5 used in this transmitter can produce a 2 W signal, and the 12AV6 can modulate it with excellent performance. The only problem is that the circuit is powered from the ac power line and has no isolation transformer. Therefore, the reader must avoid shock hazards by installing the transmitter in a plastic or wooden box, and you must be careful when using it.

Suggested uses for this transmitter include the following:

- It can be used as an experimental broadcasting station for schools, clubs, technology education, etc.
- It may be useful for providing translation services in conference rooms. The translator can use this transmitter, and people who are interested in the service can hear him via the use of a small portable AM receiver with an earphone.

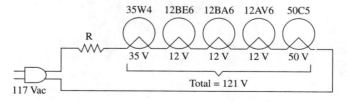

Figure 4.32 Filaments (heaters) wired in series as found in old radio receivers.

- The transmitter can be used at the lab to perform experiments with plant growth and circadian rhythms. The plants can be submitted to the high-frequency fields produced by this transmitter.

Warning: It is important for readers to be aware that this transmitter can send high-power signals over long distances, so legal restrictions on its operation must be observed. Use only a small antenna. It is best to use it in locations that are far removed from populated areas and any other places where you might interfere with the operation of licensed communication equipment.

How It Works

Two tubes are used to this project. One, the 12AV6, is an audio triode intended to operate as an audio preamplifier, and the other, a 50C5, is a power pentode intended to drive a loudspeaker through a high-impedance transformer. These tubes can be found in old radios and service shops. Many dealers have them available either in stock or via special order.

Keep in mind that, if you obtain these tubes from an old radio, other parts can also be reused, such as the tube sockets and the variable capacitor. Figure 4.33 shows a variable capacitor as used in old tube radios that can be used for our transmitter. This is a two-section variable capacitor and it is important to make sure that the mobile plates don't touch the fixed plates when the device is operated. If this occurs, the part will short out and be useless for this project.

The high-frequency oscillator is a Hartley oscillator using the 50C5. In this circuit, the coil and CV (the variable capacitor) determine the frequency.

The feedback that keeps the oscillator running is given by the tape connected to pin 6. R5 biases the control grid and R6 the cathode. Note that the tube is powered from a high voltage coming from the power supply. Keep in mind that this circuit operates at voltages up to 150 V. Therefore, you must take great care when

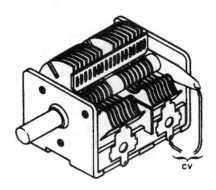

Figure 4.33   Wiring the variable capacitor.

working with the circuit when it is powered up, because there is a serious shock hazard.

The modulation signal is applied to the control grid of V2 by capacitor C4. This signal comes from V1, the preamplifier for the audio signals. The circuit is very sensitive, and medium- and high-impedance microphones can be plugged into the V1 input.

The power supply is formed by a diode that directly rectifies the ac voltage from the ac power line. The dc voltage is filtered by C1. This is a double high-voltage electrolytic capacitor. This component can also be found in old radios, but it is important to test it before using. Electrolytic capacitors wear out with time and must be tested before you use them.

Figure 4.34 shows how to test an electrolytic capacitor using a 6 V lamp and a 6 V supply. If the lamp glows in this test, indicating that the capacitor is shorted out, the capacitor cannot be used. Of course this isn't a definitive test. If the fuse opens when using this capacitor in your project, it is probable that the capacitor is the culprit. You must replace it.

Assembly

Figure 4.35 shows a schematic diagram of the tube MW transmitter. The circuit is mounted using a metallic chassis as shown in Fig. 4.36.

The holes for the tubes, transformer, wires, switch, fuse holder, and capacitor must be drilled before assembly. It is important to have these components at hand, as they can vary in appearance and dimensions. This allows the reader to make the alterations in the layout of the chassis according to the specific components used in the circuit.

After preparing the chassis, install the tube sockets, the transformer, and the variable capacitor. For the variable capacitor, because the fixed plates are connected to ground in normal operation, we need to make some modifications to use the two sections as shown in Fig. 4.37. In this figure, observe the points

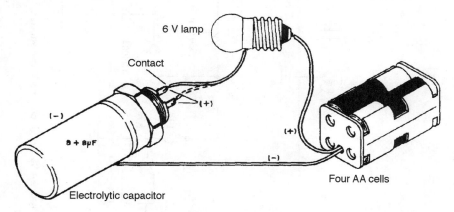

Figure 4.34   Testing an old electrolytic capacitor.

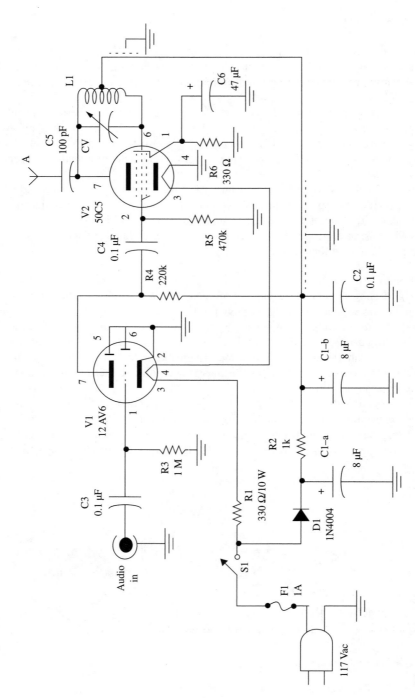

Figure 4.35 High-power AM transmitter.

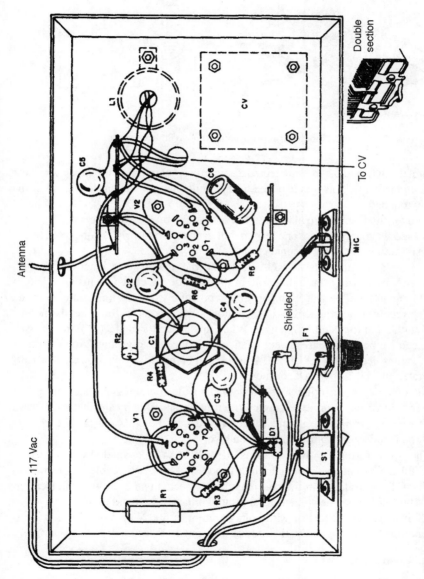

Figure 4.36 Component placement under the chassis.

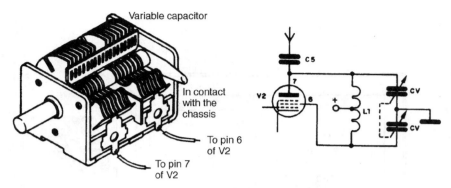

Figure 4.37   Required modifications for using the variable capacitor.

where the wires must be connected for correct operation. The capacitor must have a plastic knob, as it is exposed to high voltages and can cause a shock if directly touched.

Two terminal strips are also fixed to the chassis to receive some small components and wiring. Note that one terminal of each terminal strip is connected to the chassis, acting as a ground connection.

The audio input is via an RCA jack, and the antenna is a length of plastic-covered wire passing through a hole in the back of the chassis. This wire is 6 to 10 ft long and will be placed in convenient position when the transmitter is used.

The recommended electrolytic capacitor is the type used for "chassis mounting" with a nut to fasten it in a hole. When this capacitor is placed in the chassis, the negative pole (the case) is automatically connected to ground. Any capacitor with values between 8 + 8 and 50 + 50 µF can be used for this project. The voltage rate must be 200 WVDC or more.

Another important point to be observed in the mounting is that one of the power cord wires is connected directly to the chassis. This means that the chassis is "alive" and can cause severe shocks if touched. For this reason, the chassis must be installed in a plastic or wooden box.

The coil is formed by 40 + 40 turns of AWG 26 to 28 enameled wire in a 1 in. PVC tube. You can also use a length of broomstick or plastic cable as a form.

The CT (central tap) in the coil must be connected using a short, shielded cable. The shield must be connected to ground (chassis) to avoid hum.

C2 is a ceramic capacitor with voltage rating of 500 V or more. R1 is a wire-wound resistor with a dissipation power of about 10 W or more.

Adjustments and Use

Install a fuse in the fuse holder and place the tubes in their sockets. Observe the correct position for tube installation.

Place an MW receiver near the transmitter and tune it to a free point in the band. The antenna wire can be placed at any convenient location.

Turn the transmitter power on and wait two or three minutes while the tubes warm up. The red light produced by the electrodes in the tubes is normal, as is their high temperature. Don't touch the tubes while they are in operation, because they are very hot. If the tubes don't glow when the power is turned on, they are bad (blown) and can't be used.

Next, by adjusting CV, you will be able to tune in the signal. Plug a microphone or other audio source into the input to test the modulation.

Warning: Any device plugged into the input will be exposed to the ac power line voltage. This means that you can use only plastic microphones with insulated cables or battery-powered devices as signal sources.

If a strong hum appears when you tune in the signal, turn the ac power plug 180°. If the hum continues, test C1. This component may be defective.

To use this transmitter, you can use a mixer to transmit your voice as well as music from a tape recorder or CD player. Figure 4.38 shows an audio output transformer wired to allow the use of a small loudspeaker as a microphone.

If the audio sources are powered from the ac line, use only supplies that incorporate effective isolation transformers. Observe that the circuit also can be powered from 220 or 240 V power supplies if you incorporate an input transformer.

Parts List: Project 30

Tubes

V1   12AV6 audio triode, miniature tube

V2   50C5 audio pentode, miniature tube

Semiconductor

D1   1N4004 or 1N4007 silicon rectifier diode

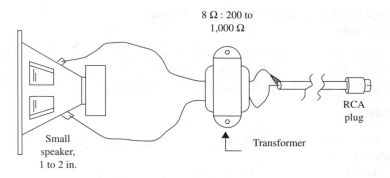

Figure 4.38   Using a small loudspeaker as a microphone.

Parts List: Project 30 (continued)

Resistors (1/8 W, 5%)

R1   330 Ω × 10 W wirewound

R2   1,000 Ω × 10 W wirewound

R3   1,000,000 Ω—brown, black, green

R4   220,000 Ω—red, red, yellow

R5   470,000 Ω—yellow, violet, yellow

R6   330 Ω—orange, orange, brown

Capacitors

C1   8 + 8 µF to 50 + 50 µF, 200 WVDC or more, double electrolytic (see text)

C2,   0.1 µF × 250 V ceramic

C3, C4 0.1 µF × 50 V ceramic

C5   100 pF/400 V ceramic

C6   47 µF/35 WVDC electrolytic

Additional Parts and Materials

L1   coil (see text)

CV   double variable capacitor (see text)

F1   1 A fuse and holder

S1   SPST toggle or slide switch

Metal chassis, power cord, sockets for the tubes, knob for CV, plastic or wooden box, microphone, etc.

Project 31: High-Power Tube CW Shortwave Transmitter

This high-power telegraphic transmitter can send signals as far as 1,000 miles if an external antenna is used. Of course, for experimental purposes (and if you're not a licensed amateur radio operator), such an antenna must not be used.

Features

- Frequency range: 3 to 7 MHz (40 to 80 m band)
- Output power: 2 to 5 W
- Power supply voltage: 117/220/240 Vac
- Produced signal: CW

If you're looking for a CW transmitter to test your skill in telegraphy or to learn Morse code this is an interesting project. As mentioned, this transmitter, if used by an amateur radio operator and equipped with an appropriate antenna, can send signals over great distances. But the principal advantage to readers who want to conduct radio transmission experiments is that this transmitter can be built using parts found in old radio and television sets.

The tubes are common in 1960s/1970s vintage TVs and radios and can be cannibalized from abandoned units. As this circuit is intended for the reader who wants projects for experimentation and education, it bears repeating that it should not be used with large or external antennas. See Chapter 1 for more information about the laws and rules of telecommunications.

How It Works

This SW transmitter can be operated in two configurations:

- It can be used as a CW (continuous wave) unit, where the transmitted signals consist of high-frequency pulses produced by a key.
- It also can serve as an AM (amplitude modulated) tone telegraphic transmitter. In this mode, the high-frequency signal is modulated by keyed pulses of a tone produced by an oscillator.

Figure 4.39 shows the wave shape of a CW signal when transmitting the letter *R* and also a tone-modulated signal transmitting the same letter (· − ·) in Morse code. The reader is free to choose either version.

In our transmitter, if the AM version is chosen, the modulation tone is produced by a relaxation oscillator using a neon lamp. In this circuit, the frequency depends on the resistor and the capacitor (RC).

When the circuit is powered on, the capacitor charges until the trigger voltage of the neon lamp is reached. When the neon lamp triggers on, the capacitor discharges through it, producing a current pulse. When the capacitor is discharged, the neon lamp turns off, and a new charging cycle begins.

The high-frequency signal (RF) to be transmitted, and which serves as the carrier for the tone signal, is produced by a Hartley oscillator using a pentode tube.

(a)

(b)

Figure 4.39 The letter *R* in Morse code: (a) CW and (b) tone modulated.

Any audio or video pentode can be used in this circuit such as the 6L6, 6V6, 6AQ5, etc. These tubes are designed for use in audio power output amplifiers but, because they have a high cutoff frequency, they can also can be used to produce high-power radio signals.

Many radio amateurs have built their own transmitters using these tubes. This configuration was very popular in the 1960s/1970s. Tube sockets can also be found in the same equipment from which you extract the tube.

Figure 4.40 shows the pinout of some tubes that can be used for this transmitter. Coil L1 and CV determine the frequency produced. CV is any variable capacitor as found in old radio sets. The use of small, plastic variable capacitors is not recommended, as they can't isolate the high voltage found in this circuit.

The output power for this circuit depends on the voltage applied to the tube. This voltage depends on the power transformer employed.

The reader can find a good transformer, suitable for this project, in many old tube TV sets and radios. The transformer must be a type with a high-voltage secondary rated to voltages between 125 and 300 V. The transformer also must have a low-voltage secondary rated to 6.3 V, which is the voltage necessary to heat the tube (filament). If the voltage applied to the tube is higher than the recommended limit, the tube can be burned out within a few moments, and the transmitter will not operate as desired.

Observe that the voltages used for this circuit are very high, which means that there is a shock hazard. Although the circuit is isolated from the ac power line by the transformer, you should exercise much care when working on it while the power is on. It is also important to use a variable capacitor that has a wide gap between the plates. If the gap is narrow, sparks can be produced, degrading the transmitter's operation.

The electrolytic capacitors used to filter the dc current must be rated to a voltage at least 80% higher than the secondary winding high voltage. For a 125 V

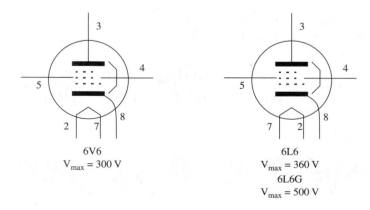

Figure 4.40 Other old tubes that are suitable for this project.

secondary winding, for instance, you should use 200 WVDC electrolytic capacitors.

In old tube-based equipment, diode tubes were used to convert the ac to dc. As we can find semiconductor diodes that offer a very low price as compared to the cost of tubes, this option was chosen for our project. The diodes recommended for all versions are the 1N4007 components.

For the antenna, you can use a piece of plastic-covered wire 3 to 10 ft long. Don't use longer antennas or external antennas, as the signal produced will be strong enough to create legal problems for you (see Chapter 1 for more information). If an external antenna is used, the signal can be picked up from the circuit by a coil of 4 or 5 turns wound over L1.

Assembly

The complete diagram of the shortwave tube transmitter is shown in Fig. 4.41. The circuit is mounted using a metallic chassis as shown in Fig. 4.42.

Locate the principal parts such as the transformer and C3/C4 before working on the chassis. These components exist in several shapes and dimensions, and the reader should design his own chassis to match the components that are on hand.

If it is difficult to find an appropriate chassis, the reader can use a metal plate by cutting, bending, and drilling holes in it. Observe that the use of a metal chassis is important, as it acts as an electromagnetic shield to prevent hum.

L1 is formed by winding 15 + 15 to 20 + 20 turns of AWG 26 to 28 enameled wire on a 1 in. PVC tube. With a coil of 20 + 20 turns, it is possible for operation between 3 and 7 MHz. With a coil of 15 + 15 turns, the frequency range will be between 5 and 10 MHz.

Note that these values are not exact, as there are component tolerances involved as well as variations in the variable capacitor. The variable capacitor, for instance, can be any type with capacitances in the range between 120 and 410 pF. Remember that the capacitor must be mounted on an insulator so it does not touch the chassis.

Figure 4.43 shows how the transmitter looks after assembly. Figure 4.44 shows how the Morse key is connected to the circuit for CW transmission. Take care when using the Morse key, as it is exposed to high voltages when the transmitter is on.

C3, C4, and C5 must be ceramic capacitors rated for high voltages; 400 V is a suitable value. C1 and C2 can be metal film types with working voltages of 100 V or more.

Testing and Using the Circuit

To test the circuit, place it near an SW receiver that has been tuned to a free point in the band. Power the transmitter on with S1 and wait two or three minutes for it to warm up. The tube will emit a weak red light when hot.

Pressing the Morse key, tune the variable capacitor until the signal is received. Adjust P1 to obtain the desired tone for modulation.

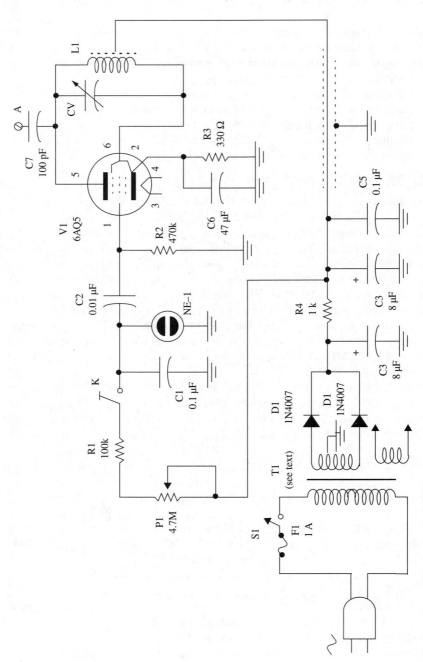

Figure 4.41   High-power CW SW transmitter.

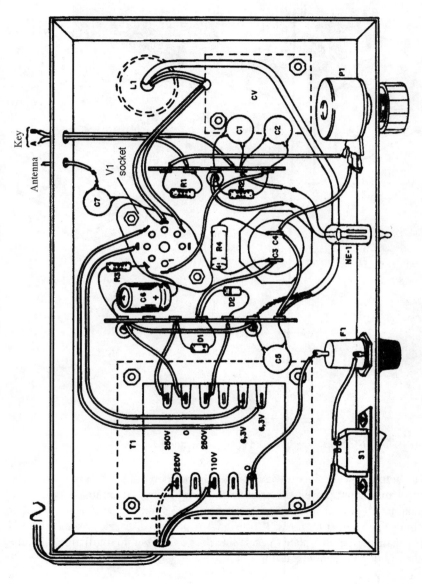

Figure 4.42 Component placement under a metal chassis.

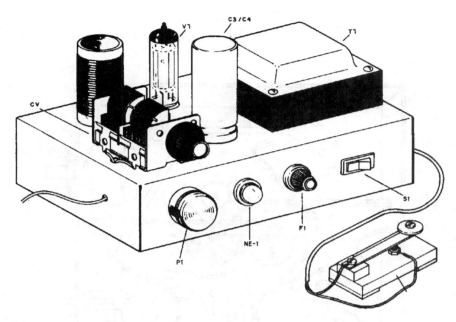

Figure 4.43   Final appearance of the transmitter.

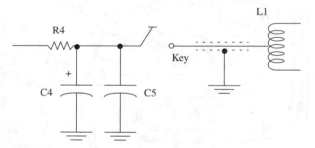

Figure 4.44   The Morse key is placed between R4 and L1.

The output power can be verified using a Hertz loop as shown in Fig. 4.45. The stronger the output power, the brighter the lamp glows. Alternatively, a fluorescent lamp will glow from the high-frequency signals if it is placed near the coil or touching the antenna terminal. If no signal is produced, check to see if there is high voltage in the tube plate (V2). If there is voltage but no oscillation, the tube is burned out and cannot be used.

If there is no voltage in the power supply output, verify the operation of the transformer and the electrolytic capacitors.

If the lamp lights up but the signal isn't picked up by the receiver, change the coil or try to find the signal in another part of the frequency band.

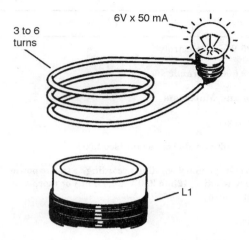

Figure 4.45   Testing the transmitter with a Hertz loop.

Parts List: Project 31

Tube

V1   6AQ5 or equivalent miniature pentode (see text) to equivalents

Semiconductors

D1, D2   1N4007 or equivalent silicon rectifier diode

Resistors (1/8 W, 5%)

R1   100,000 Ω—brown, black, yellow

R2   470,000 Ω—yellow, violet, yellow

R3   330 Ω—orange, orange, brown

R4   1,000 Ω × 5 W wirewound

P1   4,700,000 Ω potentiometer

Capacitors

C1   0.1 µF/100 V ceramic or metal film

C2   0.01 µF/200 V ceramic or metal film

C3, C4   8 µF to 32 µF/450 V or more, electrolytic (double) (see text)

C5   0.1 µF/600 V ceramic

Parts List: Project 31

C6	47 µF/25 WVDC electrolytic
C7	100 pF × 600 V ceramic

Additional Parts and Materials

L1	coil (see text)
CV	210 to 410 pF variable capacitor (see text)
T1	transformer: primary winding according to the ac power line voltage, secondary winding 125 to 250 V CT, 80 mA or more and 6.3 V × 1 A for filament heating
NE	neon lamp
F1	1 A fuse and holder
S1	SPST toggle or slide switch

Metal chassis, wire for the coil, power cord, knobs for the variable capacitor and potentiometer, Morse key, socket for the tube, etc.

5

Other Transmitters

The MW, SW, and FM bands aren't the only ones that can be used to send signals to a receiver. In addition, you can send signals to a receiver using methods other than radio waves. This chapter is dedicated to transmitters that use different technologies such as audio signals, low-frequency magnetic fields, light beams, and others that were not appropriate for previous chapters. The reader who wants to experiment with different kinds of circuits will find these to be excellent suggestions for school, technology education, and scientific research.

Project 32: Transmissions through the Earth

Can the Earth be used as a signal conductor to send messages to a remote receiver? If you don't believe that it's possible, this project certainly will change your mind.

Features

- Transmitter: any 10 to 100 W audio amplifier
- Range: up to 1 mile (depending on the amplifier power and the antenna)
- Frequency range: audio (16 to 16,000 Hz)
- Technology: low-frequency current field

During World War II, when German troops occupied France, French radio amateurs were prohibited from using their radio systems. But communication was very important to the "resistance," as they needed to disseminate important messages. The French radio amateurs soon discovered that, even if the "air" couldn't be used to send messages, the Earth could provide an excellent substitute. It was enough to place two electrodes into the ground and connect them to the output of any audio amplifier. The current field produced by this system could be picked up by two distant electrodes, with the transmitted signal reproduced in an earphone. Using amplifiers with powers in the range of 10 to 40 W, it was possible to send messages over distances up to several miles.

This very simple transmission system has many drawbacks, as the signals are not tuned, and the electrodes must be placed as far as possible from each other.

However, the reader can use it for conducting experiments or to build a short-range communication system.

The principal advantage is that the signal can't be picked up by common radio receivers and causes no interference to other communication systems. If you use two audio amplifiers, a bilateral communication system easily can be built.

How It Works

The system is based on "current fields." If we place two electrodes in a homogenous conductor such as the Earth, as shown in Fig. 5.1, and apply a voltage between them, a current field will be produced.

The current flowing between the electrodes will be distributed by field lines (as in a magnetic field). If we place two electrodes across these lines, a voltage will be detected with a value that depends on their separation along the particular line.

Notice that, between different lines, it is possible to draw an imaginary line along which all of the voltages are equal. This is called the *equipotential surface*, as it is not really a line but a surface (as shown in Fig. 5.1).

The reader, from this explanation, can easily see that, if we place two electrodes at a convenient location along the current field, it is possible to pick up the signal. Even if the electrodes are placed very far away from the transmitter (current source), a small voltage can be picked up.

In theory, the electrodes can be placed at distances as far as infinity, but the practical sensitivity of the receiver will limit this distance to a few hundred feet up to several miles.

At a great distance from the transmitter, the voltage will be overwhelmed by other induced signals existing in the ground, such as hum, atmospheric discharges, and many other noise sources.

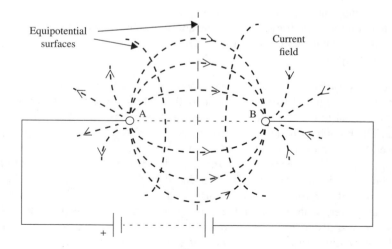

Figure 5.1 Current field between electrodes A and B.

For direct current and low-frequency signals, the principle is valid as described. But as frequencies rise, other factors begin to act on the current fields. For high-frequency signals, attenuation increases, and wave propagation in the air becomes more important than the current fields in the Earth.

To obtain the best performance from a system like this, it is important to consider some technical facts. First of all, if you plug the output of any amplifier into the ground, the audio-frequency signal must be completely transferred, which means that you must establish a correct impedance match. The ground's impedance depends on several factors, including the distance between the electrodes, the electrodes' size, and the type of soil in which the electrodes are placed. This impedance can range from several ohms to hundreds of ohms.

The best away to match the amplifier output impedance with the soil impedance is via the use of a transformer as shown in Fig. 5.2. A transformer with several taps can be used to find the best impedance match for an Earth communication system. The electrodes used as "antennas" must be as far apart as possible. Distances between 15 and 60 ft are recommended.

The simplest receiver is a high-impedance earphone plugged into two electrodes as shown in Fig. 5.3. Note in this figure that two relative positions for the receiver electrodes as related to the transmitter electrodes are possible. But the best sensitivity can be achieved if the reader uses an amplifier with the receiver. A loudspeaker can be driven if a medium- to high-power amplifier is used.

For a bilateral system, as we will describe herein, the same amplifier used to transmit the signals can be "inverted" to act as a receiver.

Recall that this is not a tuned system, as the audio signal is its own carrier, so the frequency range is spread between 16 and 16,000 Hz. As a result, if two or more transmitters are operating simultaneously in the same area, it will not be possible to separate the signals.

Also, atmospheric discharges (lightning), noise produced by industrial machinery, and many other noises will be superimposed on the signals, and it is not

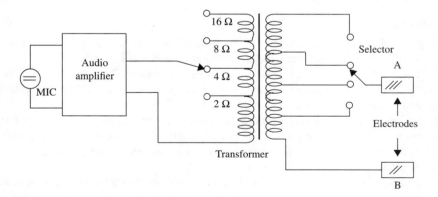

Figure 5.2 Impedance matching using a transformer.

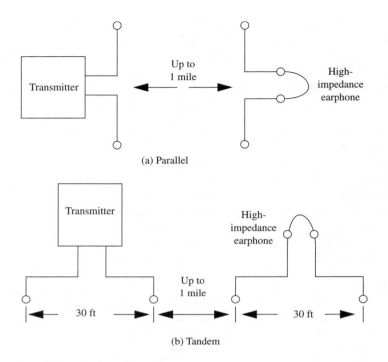

Figure 5.3   Electrode placement.

possible to avoid their influence. You can use audio filters to improve the system's performance. You can increase the system's range if you just use an audio tone to send telegraphic messages in Morse code.

Summarizing, our project consists of an audio amplifier with additional circuits to transfer the signals to the ground and also to pick up signals from the ground at the receiving end.

Assembly

Figure 5.4 is a complete diagram of the "through the Earth" communication system. The printed circuit board to the amplifier is shown in Fig. 5.5.

The TDA2002 is an audio amplifier integrated circuit. The output power is about 5 W, which is enough to communicate over distances up to 600 ft. The integrated circuit must be mounted on a heat sink.

For the input, we can connect any high-impedance microphone and adjust P1 to get the highest output power. The power supply uses a transformer with a primary winding rated to the ac power line and a secondary winding that produces 12 V CT with current of 1 A or more.

T2 is any transformer with a low-impedance winding with taps and a high-impedance winding with taps. You can experiment with a power transformer with a primary winding rated to 117/220 Vac and a secondary winding to 6 V to 12 V

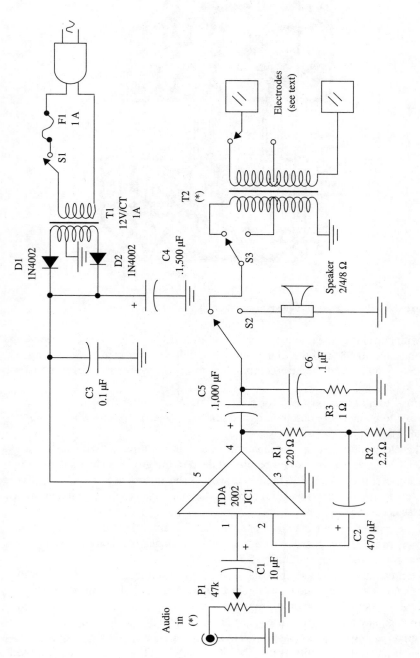

Figure 5.4   Through-the-Earth communication system.

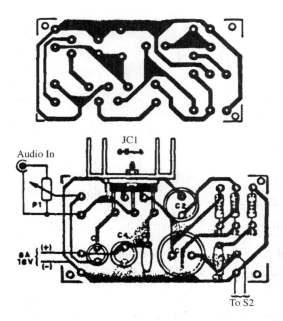

Figure 5.5   Printed circuit board for the amplifier stage.

CT with currents in the range between 500 mA and 1 A. The low-voltage winding is connected to S3, and the high voltage winding is connected to S4. Using S3 and S4, you can find the best combination to obtain maximum energy transference to the Earth.

Any 4 to 8 Ω speaker rated for power of 10 W or more can be used. It is also possible to connect the output to a medium-impedance earphone when using the circuit as a receiver.

The electrodes used as antennas are metal bars or plates with dimensions shown in Fig. 5.6. They should be placed deep in the ground. The reader can increase the soil conductivity, thereby lowering the impedance, by adding water and some salt. The wires used to connect the electrodes to the circuit can be as long as necessary.

If the reader doesn't intend to use an identical amplifier for the receiver (for field operation), it is possible to use a high-impedance earphone (crystal or ceramic) connected as shown in Fig. 5.7.

We must alert the reader that many earphones used in CD players and Walkmans® are low-impedance types rather than the required high-impedance types, so they will not function with this circuit.

Another suggestion is to use a small, two-transistor audio amplifier that is powered from two AA cells as shown in Fig. 5.8. Other low-power amplifiers using such integrated circuits as the LM386, TBA820, TDA7052, etc. can be used in a portable receiver for this system.

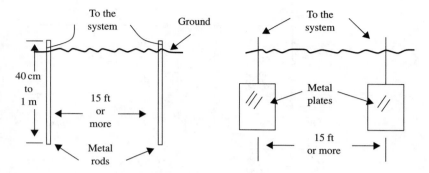

Figure 5.6 Installing the electrodes.

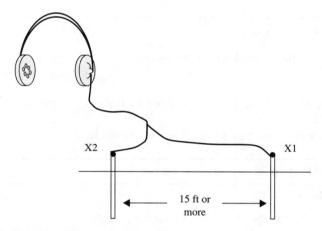

Figure 5.7 Using headphones as the receiver.

Adjustments and Use

First, test the transmitter amplifier by speaking into the microphone with the loudspeaker connected to the output. Adjust P1 to achieve the highest output level without distortion.

Next, switch S2 to the transmit position, placing the receiver at a distance from the transmitter. Speak into the microphone and see if your voice can be heard in the receiver. Find the best combination of S3 and S4. You may also be able to improve performance if you use several electrodes that can be engaged by a switch.

Other Suggestions

The reader can conduct research on ways to improve this system. Some suggested experiments follow:

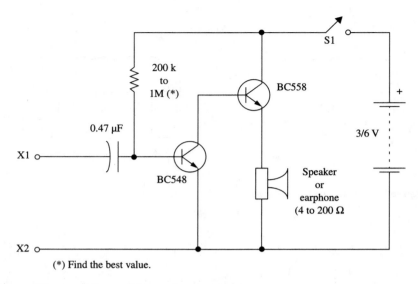

(*) Find the best value.

Figure 5.8   An audio amplifier used as a receiver.

- Try using an ultrasonic signal modulated by the audio signal and receiving it with a PLL filter.
- Try using a phase modulated signal as used by modems.

Parts List: Project 32

Semiconductors

IC1	TDA2002 audio amplifier integrated circuit
D1, D1	1N4002 or equivalent silicon rectifier diodes

Resistors (1/8 W, 5%)

R1	220 Ω—red, red, brown
R1	2,2 Ω—red, red, gold
P1	47,000 Ω—potentiometer

Capacitors

C1	10 µF/16 WVDC electrolytic
C1	470 µF/12 WVDC electrolytic
C3	0.1 µF ceramic
C4	1,500 µF/25 WVDC electrolytic

Parts List: Project 32 (continued)

C5	1,000 µF/25 WVDC electrolytic
C6	0.1 µF ceramic or metal film

Additional Parts and Materials

SPKR	4 to 8 Ω—4 inch loudspeaker
S1	SPST toggle or slide switch
S2, S3, S4	DPDT toggle or slide switch
T1	transformer, 117 Vac primary winding, 12 V CT secondary winding
T1	any high-impedance to low-impedance transformer (see text)
F1	1 A fuse and holder
X1, X1	electrodes (see text)

Printed circuit board, plastic box, power cord, knobs for the potentiometer, wires, microphone (see text), solder, etc.

Project 33: Small UHF Transmitter

This transmitter can send signals in the range between 470 and 650 MHz. The project can be used for short-range communications such as wireless microphones and spy transmitters.

Features

- Power supply voltage: 3 to 6 Vdc
- Range: 150 to 1,500 ft (see text)
- Number of transistors: 1
- Frequency range: 470 to 650 MHz
- Current drain: 5 to 20 mA (typical)

One of the most important advantages of using a transmitter such as this one is that it doesn't need an antenna. If an antenna is used to increase performance, it can be a very short one. Of course, the principal disadvantage is that you need a receiver that can tune in the UHF signals produced by this small transmitter.

The suggested applications for this circuit are:

- Short-range communications
- Use as a wireless microphone
- Espionage, to listen to conversations in adjacent rooms

Building the transmitter is very easy, as it uses common components. The critical component is the coil, but the reader need not be intimidated by it, as it is incorporated in the printed circuit board. It is also recommended that you use a fiber printed circuit board, as high-frequency circuits don't work well with other material types.

How It Works

The project is based on a common one-transistor oscillator in the common base configuration as described for many other projects in this book. L1 and CV determine the operational frequency. CV must be adjusted to a free point of the UHF band that corresponds to the receiver frequency.

As the coil is a very low-inductance component, it is incorporated in the printed circuit board instead of installed as a separate component. Thus, the half-turn coil L1 isn't a real component in this project but a "virtual" component, as the reader can see.

The feedback that keeps the oscillator running is given by capacitor C3. This also is a homemade component. This low-capacitance device can be made using two pieces of common plastic-covered wire as shown in Fig. 5.9. A length of wire as shown in the figure can form a 0.5 to 1 pF capacitor as required for this project.

The reader can use any transistor suitable for UHF operation such as the ones found in TV tuners and other sources. Types such as the BF689, BF970, BF979, and others can be evaluated. Figure 5.10 shows the terminal layout for some of these transistors.

The modulation signal comes from an electret, two-terminal microphone. This component is very small and has an excellent sensitivity due an internal FET.

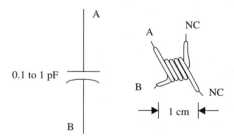

Figure 5.9   A homemade low-capacitance capacitor.

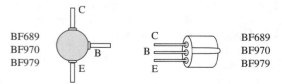

Figure 5.10   Different case layouts for the same transistors.

Resistor R1 is necessary just to bias the internal transistor, and its value depends on the power supply voltage. Current drain also depends on the power supply voltage. Values between 5 and 20 mA are typical for this project.

There are several suitable power supply options. Button cells up to 9 V batteries are acceptable as long as you change the values of some components as
shown in the table below.

Power supply	R1	R2	R3	R4
3 V	2.2k	4.7k	5.6k	47
6 V	4.7k	6.8k	8.2k	56
9 V	10k	10k	12k	100

Assembly

A schematic diagram of the UHF transmitter is shown in Fig. 5.11. The circuit
must be mounted on a printed circuit board as shown in Fig. 5.12. Special care
must be taken with the coil, as any failure or other problem can affect the transmitter's performance.

All of the capacitors must be ceramic types except C3, which can be a home-
made version as described. The trimmer capacitor is a low-capacitance type, with
its lowest capacitance at about 1 pF. If trimmers with higher capacitance ratings
are used, that will lower the transmitter frequency range, and it will be impossible
to pick up the signal at the expected point. Notice in Fig. 5.12 that the lowest capacitance is obtained when the plates are in the position that provides widest separation.

An antenna is optional. If installed, it can be a piece of rigid plastic-covered
wire no more than 4 in. long. The circuit can be installed into a small plastic box
that includes the batteries.

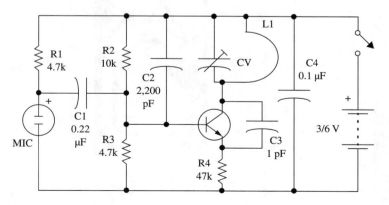

Figure 5.11 Small UHF transmitter.

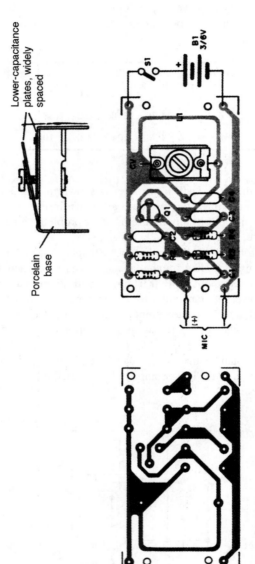

Figure 5.12   Printed circuit board for the UHF transmitter.

Adjustments and Use

The reader must have a UHF receiver (scanner) that can tune frequencies in the range between 300 and 800 MHz. Tune it in a point between 470 and 650 MHz. A UHF television set can also be used. With a trim adjustment, it is possible to pick up the transmitter's signal in an audio channel located between channels 14 and 20.

Power the transmitter on and, using a nonmetallic tool, adjust the trimmer until you tune in the signal. As the circuit is very sensitive, any movement, or the presence of objects (e.g., your hand), can detune it. Install it in the box and make new adjustments. When using the transmitter, don't shake it or make quick movements, as the circuit again can be detuned.

Note that, if the circuit cannot be tuned to the desired frequency, you can adjust C3 by reducing the wire length or use a different trimmer capacitor (CV). When using the circuit as a "bug," install it away from metallic objects that can affect signal propagation.

Parts List: Project 33

Semiconductors

Q1   BF689 or equivalent UHF NPN silicon transistor (see text)

Resistors (1/8W, 5%)*

R1   2,200 Ω—red, red, red

R1   4,700 Ω—yellow, violet, red

R3   5,600 Ω—green, blue, red

R4   47 Ω—yellow, violet, red

Capacitors

C1, C4   0.1 μF ceramic

C1   2,200 pF ceramic

C3   1 pF ceramic or home made (see text)

CV   1–10 pF trimmer (see text)

Additional Parts and Materials

MIC   electret microphone, two terminals

L1   coil, printed circuit (see text)

S1   SPST toggle or slide switch

B1   3 to 9 V, cells, battery, etc. (see text)

Printed circuit board, battery holder or clip, plastic box, solder, wires, etc.

---

*Note: resistor values are for a 6 V version. See text for other voltages.

Project 34: Experimental Digital Transmitter

This is a *one-component transmitter* to be used in demonstrations, as a signal injector, or to show how digital circuits can produce radio frequency signals.

Features

- Range: 1 to 3 ft
- Frequency range: 100 kHz to 4 MHz
- Power supply voltage: 3 to 12 Vdc
- Number of components: 1

This experimental transmitter uses only one CMOS integrated circuit and can produce radio frequency noise in a wide range of the LW, MW, and SW bands. Its signals can be tuned in by any AM or SW receiver in the form of a noise that proves that digital circuits can produce radio signals.

Other uses to this transmitter are

- Short-range telegraphic communications, sending signals through a wall
- As a signal injector to test and adjust radio receivers in the MW and SW bands
- To produce interference in radio receivers for counterespionage

The circuit can be powered from power supplies ranging from 3 to 12 V, and all the reader needs to put this device on the air is a length of plastic-covered wire as an antenna and, optionally, a length of the same wire as a ground connection.

How It Works

The inverters found in a 4049 or 4069 CMOS integrated circuit are wired as a ring oscillator. The propagation time of the signal through each inverter determines the time of a complete cycle of the resultant signal and therefore the frequency.

Notice that each stage of the circuit inverts the signal so, to obtain oscillations, an odd number of them must be used. The ideal number to keep the circuit in a state of oscillation is three or five. This means that it is not necessary to use external elements to set the frequency or to generate the feedback that is required to put the circuit in oscillation.

As the propagation time of a signal in a CMOS circuit depends on the power supply voltage, this element can be used to tune the signal. Thus, we have a fixed-frequency version in which the signal frequency is fixed by the power supply voltage, and a variable-frequency version that includes elements to change this parameter. As the circuit produces a square wave that is rich in harmonics, it is possible to tune to it at many points in the radio band. Therefore, if it is adjusted to a frequency of about 100 kHz, the signals can be picked up at points separated by 100 kHz from this frequency to up to 10 MHz and beyond. The reader can locate the frequency at which the signal power is highest and use it for the experiments.

Assembly

Figure 5.13 shows three basic versions of this transmitter. The first circuit (top) is a fixed-frequency unit in which the power supply voltage determines the signal frequency in the range of 100 kHz to 4 MHz. The second and third circuits employ added elements that can be used to change the frequency without changing the power supply voltage. All versions can be mounted on a solderless board or breadboard, as they are experimental.

The output power produced by this circuit is very low (a few milliwatts), but the transmission can be picked up in a receiver placed several feet away from it. The reader can wire a Morse key in series with the power supply for sending CW signals.

The operational range with a 6 V supply is between 1 and 3 MHz. An antenna can be connected to the circuit to increase the output power. A piece of plastic-covered wire 3 to 10 ft long makes a good antenna for this circuit.

Using the Circuit

There are no adjustments required for this circuit except in the variable frequency versions. Place any receiver tuned to a free point in the MW or SW band and turn on the transmitter. Then locate the signal in the band, adjusting the power supply

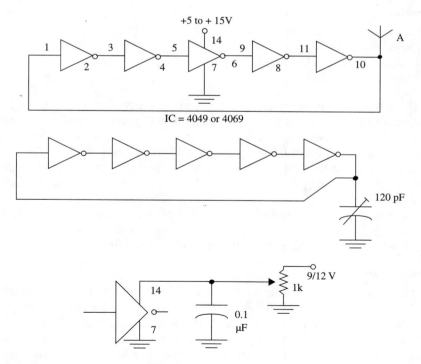

Figure 5.13   Digital transmitter.

voltage if necessary. If the circuit is used in science demonstrations, you can use it to illustrate how electromagnetic waves are generated.

Parts List: Project 34

Semiconductors

IC1   4049 or 4069 CMOS integrated circuit

Miscellaneous

Solderless board, wires, power supply, antenna, additional components for the variable-frequency versions

---

Project 35: CD Player/Walkman® FM Transmitter

This circuit can send music from a portable CD player or Walkman to any FM receiver placed in distances up to 100 ft.

Features

- Power supply voltage: 3 Vdc
- Current drain: 10 to 20 mA (typical)
- Frequency range: 88 to 108 MHz
- Transmission mode: FM mono
- Range: 50 to 100 ft

This transmitter can be installed in a small box (no larger than a cigarette pack) and used to transmit the music from a portable CD player or Walkman (tape player) to an FM receiver in your home or car. The transmission is monophonic, but the audio has excellent fidelity, as the two channel's signals are mixed. When traveling, you can use this device to send music to the car's more powerful sound system, allowing it to be heard through the loudspeakers rather than headphones or a small built-in speaker.

The circuit can also be used as a very low-power radio station to distribute programs from a tape recorder to a number of FM receivers placed nearby. The circuit is a low-power device, so the batteries will last a long time.

How It Works

The audio signals coming from the CD player or tape recorder output are mixed by C1 and C2 and applied to a modulation control. The modulation control is a trimmer potentiometer and is used to fix the level of the modulation signal.

The exact adjustment point for this component is determined by the audio source (CD player or tape recorder) volume. Excessive signals can saturate the circuit, causing overmodulation and severe distortion, as explained in Chapter 1.

The transmitter is formed by a high-frequency oscillator that uses one transistor. The frequency is determined by L1 and CV. CV must be adjusted to a free point in the FM band.

The feedback that keeps the oscillator running is provided by C5. R1 and R2 bias the transistor, which is wired in the common base configuration. This circuit is modulated by the audio signals coming from the audio source.

The high-frequency signal is applied to an antenna for transmission. The antenna is a length of rigid plastic-covered wire 4 to 10 in. long.

Two AA cells form the power supply, for more power, you can use four AA cells without changing the circuit or any component values. If the circuit is intended for use in the car, a step-down converter can be used with the 12 V automotive battery. A circuit based on the 7806 IC can be used for this task.

Assembly

Figure 5.14 is a schematic diagram of the CD player/Walkman® transmitter. The circuit can be assembled using a printed circuit board as shown in Fig. 5.15.

If you are a beginner or a student who does not have the resources to etch your own printed circuit board, you can build the terminal strip version shown in Fig. 5.16. This version is more critical, as the component's terminals and wires must be kept as short as possible. Long wires or terminals can induce the circuit to generate spurious oscillations and instabilities.

L1 is formed with four turns of AWG 18 to 22 enameled or common plastic-covered wire using a pencil as a reference (diameter of about 1 cm). The antenna can be connected to a tap, with the proper one determined experimentally as de-

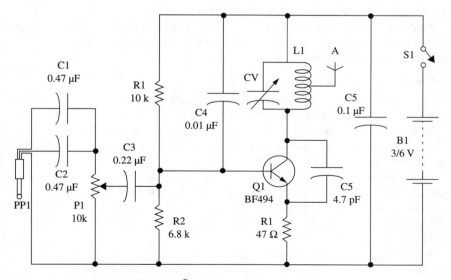

Figure 5.14   CD player/Walkman® FM transmitter.

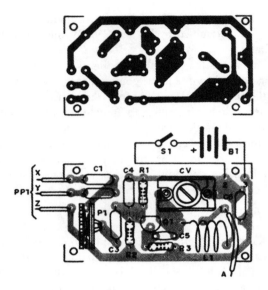

Figure 5.15 Printed circuit board for the CD player transmitter.

scribed for many other transmitters in this book. The antenna is a piece of rigid plastic-covered wire 4 to 10 in. long.

Any trimmer capacitor with a capacitance range between 2–20 and 4–40 pF can be used. All capacitors are ceramic types.

The reader can change the transistor to a 2N2218 and power the circuit from a 12 V supply, thereby increasing the output power.

Figure 5.17 shows how a stereo plug is wired to be used in this circuit. Keep the wires short, and take care not to allow any wires to touch each other, which can cause shorts that damage the source circuitry.

Adjustments and Use

Place an FM receiver near the transmitter and tune it to a free point in the FM band. Plug a Walkman® or CD player into the transmitter. The signal is taken from the earphone output using the plug arrangement shown in Fig. 5.17. The sound source must be adjusted to a low volume point to avoid overmodulating the transmitter, which causes distortion.

Close trimpot P1 and power on the transmitter. Adjust CV to tune the signal to the receiver. Carefully open P1 until the sound from the audio source is heard with best volume and without distortion. Whenever you use the transmitter for this purpose, the source volume should be the same. This is important to avoid having to readjust P1.

If the sound reproduction tends to have too much treble, increase the values of C1 and C2. When using the transmitter in the car, position the antenna for the best reception without interference or noise.

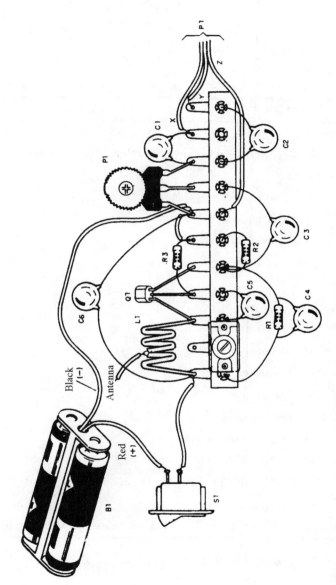

Figure 5.16   Printed circuit board for the CD player transmitter.

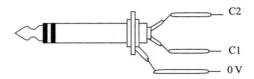

Figure 5.17   Adapting a stereo plug.

Parts List: Project 35

Semiconductors

Q1   BF494 or equivalent high-frequency NPN silicon transistor

Resistors (1/8W, 5%)

R1   10,000 Ω—brown, black, orange

R1   6,800 Ω—blue, gray, red

R3   47 Ω—yellow, violet, black

P1   10,000 Ω—trimmer potentiometer

Capacitors

C1, C2, C6   0.1 µF ceramic

C3   0.22 µF ceramic or metal film

C4   0.01 µF ceramic

C5   4.7 pF ceramic

CV   trimmer (see text)

Additional Parts and Materials

S1   SPST toggle or slide switch

B1   3 V, two AA cells (see text)

PP1   stereo plug according to the signal source (earphone output)

L1   coil (see text)

Printed circuit board, battery holder, plastic box, antenna, wires, solder, etc.

Project 36: Spyphone: The Super-Micro FM Transmitter

In this project, a high-gain, ultrasensitive stage picks up conversations and other sounds in a room, transmitting them to a remote receiver placed at a distance up to 200 ft. This is "spyphone," a transmitter used by "professional" spies.

Features

- Three-stage high-gain audio amplifier
- Frequency range: 88 to 108 MHz
- Power supply voltage: 6 VDC (four AA cells)
- Range: up to 200 ft
- Number of transistors: 4
- Current drain: 5 mA (typical)

This project adds an ultra-high-gain audio amplifier to an FM transmitter. As a result, very weak sounds picked up by an electret microphone can be transmitted and heard in a remote receiver.

The high-sensitivity characteristic and the high stability due to a zener regulation stage make this transmitter ideal for secret missions. Hidden in a room, it can send any conversation, even a murmur, to a receiver that is plugged into a tape recorder or an earphone.

Our suggestion is install the circuit inside a hollowed-out book (Fig. 5.18) and place it on a shelf near the conversations to be overheard. In this way, it will not be noticed. But you can hide the circuit in other objects such as flower pots, conceal it behind tables, place it among other books, or hide it anywhere your imagination prescribes. The low current drain provides a battery life of several weeks.

How It Works

The audio amplifier is formed by three transistorized stages. The first transistor picks up the signal from an electret microphone by C1. The transistor wired in

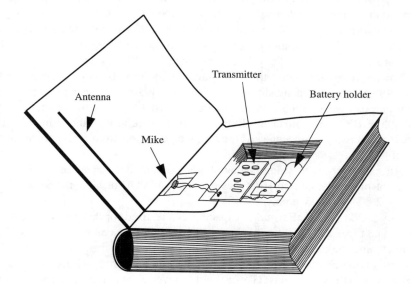

Figure 5.18   Concealing the circuit in a hollowed-out book.

the common emitter configuration is biased by R2. The signal amplified by Q1 is applied to the second stage by C3. The second stage is formed by Q2, which is biased by R4.

From this transistor collector, the signal is applied to the third stage via C5. R6, placed in parallel with this capacitor, biases the third transistor's base.

All three transistors are powered from a regulated power supply. The voltage is fixed by the zener diode at 2.4 V. Resistor R8 reduces the current flow through the zener diode to a compatible level. Filtering of the supply to this stage is provided by C4. This regulation stage is important, as it provides stability to the circuit even when the cell's voltage falls during use.

The high-frequency stage is formed by Q4 and the components placed around it. L1 and CV determine the oscillation frequency. CV must be adjusted to a free point in the FM band between 88 and 108 MHz.

To avoid saturating the circuit, the modulation signal is attenuated by R10. R11 and R12 bias the oscillator transistor, and C8 provides the necessary feedback to keep this circuit running.

Note that the antenna can be connected to any point between the collector and a tap in the coil. The reader, through experimentation, must determine the best point to connect the antenna. C9 and C10 are used to decouple the power supply.

Assembly

A complete diagram of the spyphone is shown in Fig. 5.19. All the components, except the battery holder and the electret microphone, are mounted on a printed circuit board as shown in Fig. 5.20. Notice that the resistors are mounted in a vertical position, and the electrolytic capacitors are parallel-terminal types. This is important to achieve a compact mounting for the circuit as shown in the figure.

The coil is formed by four turns of AWG 18 to 22 enameled wire or common plastic-covered wire wound on a pencil as a reference (approx. 1 cm dia.). The antenna is a piece of solid plastic-covered wire 6 to 20 in. long.

For the transistor, there are many suitable equivalents. For instance, for the BC548, the reader can substitute a BC547, BC549, or any other high-gain, general-purpose NPN silicon transistor. For the BF494, we can recommend the BF495 or BF254 as equivalent. But you must be careful with the terminal placement when using equivalents, as they may be different from arrangement of the originals.

All capacitors with values below 1 µF are ceramic. Capacitors with values above 1 µF can be electrolytic types rated to 6 WVDC or higher. The zener is a 2.4 or 2.7 V, 400 mW type.

Any 2–20 to 4–40 pF trimmer capacitor can be used for this project. Plastic or porcelain types are suitable. It is important to have these components in hand before etching the printed circuit board, as modifications in the layout may be necessary to match the device's terminal placement. When wiring the electret microphone, the reader must take care not to reverse the connections, as it is a polarized component.

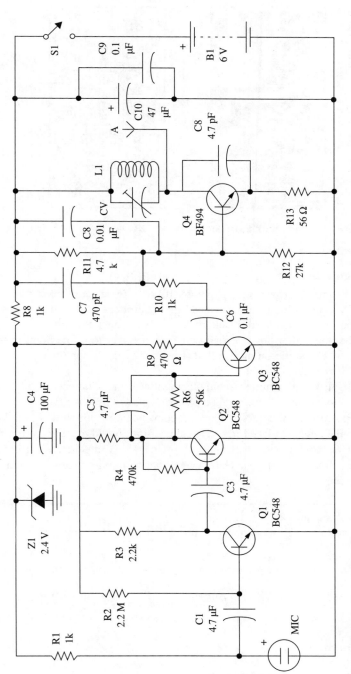

Figure 5.19   Spyphone.

 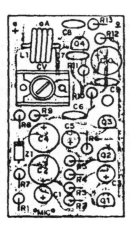

Figure 5.20   Printed circuit board for spyphone.

Adjustments and Use

Place an FM receiver near the spyphone. If possible, use a receiver with an earphone. If you use a loudspeaker instead of an earphone, the high-gain audio circuit can produce acoustic feedback, making adjustments difficult.

Turn the transmitter power on and tune the receiver to a free point in the FM band. Adjust CV until you pick up the strongest signal. When tuned, the ambient sounds will be clearly audible.

When using the transmitter, avoid placing it near metal objects. If it is installed inside a hollow book, the unit can be placed as shown in Fig. 5.21.

When positioning the book to pick up conversations, you will get the best results with the antenna placed in a vertical position. Avoid placing the transmitter on a table, as moving objects in contact with the table can generate noise that will interfere with your reception. The receiver can be plugged into the input of any tape recorder.

Parts List: Project 36

Semiconductors

Q1, Q2, Q3   BC548 or equivalent general-purpose NPN silicon transistor

Q4   BF494 or BF495 high-frequency NPN silicon transistor

Z1   2V4 (2.4 V) × 400 mW zener diode

Resistors (1/8W, 5%)

R1, R5, R8   1,000 Ω—brown, black, red

Parts List: Project 36 (continued)

R1	2,200,000 Ω—red, red, green
R3	2,200 Ω—red, red, red
R4	470,000 Ω—yellow, violet, yellow
R6	56,000 Ω—green, blue, orange
R7	560 Ω—green, blue, brown
R9	470 Ω—yellow, violet, brown
R10	100,000 Ω—brown, black, yellow
R11	47,000 Ω—yellow, violet, orange
R11	27,000 Ω—red, violet, orange
R13	56 Ω—green, blue, black

Capacitors

C1, C3, C5	4,7 µF/6 WVDC electrolytic
C2, C4	100 µF/6 WVDC electrolytic
C6, C9	0.1 µF ceramic
C7	470 pF ceramic
C8	4.7 pF ceramic
C10	47 µF/6 WVDC electrolytic
CV	trimmer (see text)

Additional Parts and Materials

L1	coil (see text)
MIC	electret microphone - two terminals
S1	SPST toggle or slide, miniature switch
B1	6 V, four AA cells

Printed circuit board, battery holder, plastic box or hollow book, solder, wires, etc.

Figure 5.21 Conversations can be picked up by the spyphone placed among real books.

Project 37: Through-the-Wall Communication System

You can send telegraphic audio signals from one room to another, through the walls, using a magnetic loop. This simple circuit can be used for short range communications.

Features

- Power supply voltage: 6 Vdc
- Range: 3 to 4 ft
- Frequency range: audio (400 Hz to 5 kHz)
- System used: magnetic loop

An important application for this device is in science fairs or as a project in technology education involving telecommunications. But the basic application suggested for this project can also be used to communicate with a friend in an adjacent room.

The circuit is a magnetic loop transmitter/receiver operating with signals sent by the magnetic field and produced by a coil. The circuit is intended for very short-range communications. Its principal advantage is that the signals can pass through solid objects such as walls.

The power supply comes from four AA cells, and each unit can both receive and transmit signals. Using a pair of these devices, the reader can achieve two-way communications. Figure 5.22 shows how the circuit can be used as a room-to-room intercom.

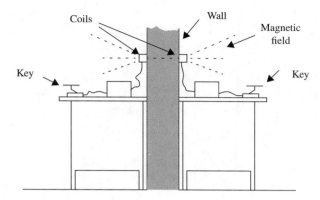

Figure 5.22   Using the system in a "through-the-wall" communicator.

How It Works

The circuit is formed by a two-stage transistor amplifier that can be used both as an amplifier and as a low-frequency oscillator, depending on the position of S1. When S1 is in the receiver position, the signals picked up by L1 are applied to the base of the first transistor and amplified. The amplified signals appear in the second transistor's emitter from which they are applied to a small loudspeaker.

When S1 is in the transmitter position, the circuit acts as an audio oscillator where the frequency is determined by C2 and the adjustment of P1. The audio signal is applied to a coil (L2), producing the magnetic field.

Note that, for best performance, the magnetic field lines produced by the transmitter's coil should intercept the receiver's coil as shown in Fig. 5.23.

The current drain is higher when the device is transmitting. In the receiving state, the current drain is determined by R1. The reader must conduct experiments to find the best value for this component. Values between 470 kΩ and 2.2 MΩ can be tested.

Assembly

Figure 5.24 shows the complete diagram of one unit, including a receiver and a transmitter circuit. As the circuit is experimental and intended for beginners, the

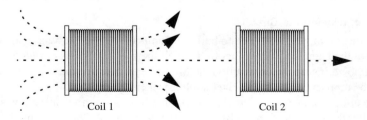

Figure 5.23   Magnetic field lines produced by coil 1 intercepted by coil 2.

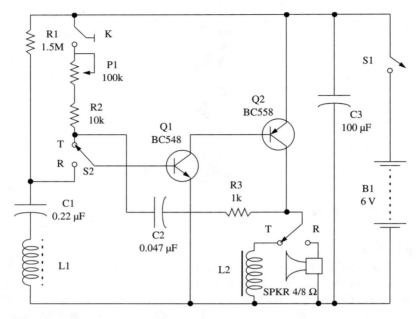

Figure 5.24 Through-the-wall communicator.

components can be mounted on a terminal strip as shown in Fig. 5.25. The coils are a power transformer's windings. The metal core is removed, and only the windings are used.

For the receiving unit (L1), the primary 117 Vac winding of any small power transformer can be used. The low voltage winding (6 to 12 V × 150 to 500 mA) can be used.

Any small 4 to 8 Ω loudspeaker or earphone can be used to reproduce the signals. The dimensions of the plastic box will be determined largely by the loudspeaker's dimensions.

The Morse key can be a homemade version built with a wooden base and metal plates as shown in Fig. 5.26. P1 can be a common potentiometer adjusted to determine the transmitted tone or, if the reader prefers, a trimmer potentiometer. Figure 5.27 shows the device installed in a plastic box and ready to be used.

Testing and Using the Circuit

To test the circuit, you will need to build two units of this circuit, with one placed to receive the signals and the other to transmit them as suggested in Fig. 5.28.

After powering up the units, press the Morse key in the transmitter. The signals should be reproduced in the receiver's loudspeaker or earphone.

If the circuit cannot send signals more than a few inches, indicating low sensitivity, alter the value of R1 and also reduce the value of C2. Placing a small ferrite core inside the coils will increase performance. Try this solution.

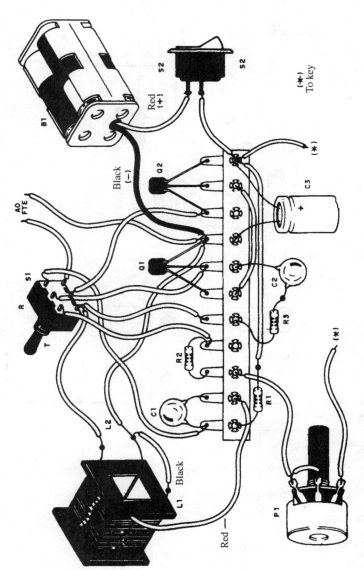

Figure 5.25 Mounting the circuit on a terminal strip.

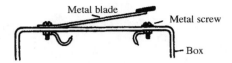

Figure 5.26   A homemade Morse key.

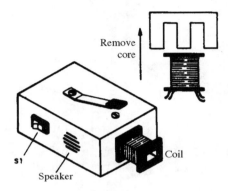

Figure 5.27   The circuit installed in a plastic box.

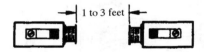

Figure 5.28   Placement of devices for initial test.

Parts List: Project 37

Semiconductors

Q1   BC548 or equivalent general-purpose NPN silicon transistor

Q1   BC558 or equivalent general-purpose PNP silicon transistor

Resistors (1/8W, 5%)

R1   1,500,000 Ω—brown, green, green

R2   10,000 Ω—brown, black, orange

R3   1,000 Ω—brown, black, red

P1   100,000 or 220,000 Ω—potentiometer

Parts List: Project 37 (continued)

Capacitors

C1	0.1 µF ceramic or metal film
C2	0.022 µF ceramic or metal film
C3	100 µF/12 WVDC electrolytic

Additional Parts and Materials

S1	DPDT toggle or slide switch
S2	SPST toggle or slide switch
B1	6 V, four AA cells
L1, L2	coils (see text)
SPKR	4 or 8 Ω—small loudspeaker (2 in.)
M1	Morse key

Terminal strip, plastic box, battery holder, wires, solder, etc.

Project 38: Laser Beam Communication System

You can use a laser pointer to build an experimental communication system. This project, to be developed by the reader, is ideal for science fairs or as a project in technology education. It is also of interest to readers who work in various scientific fields.

Features

- Range: 300 to 1,000 ft (depending on the laser)
- Modulation: amplitude
- Transmitted signals: audio (200 to 10,000 Hz)

You can send voice or music from any source (e.g., a CD player or tape recorder) over distances up to 1,000 ft using a laser beam. This experimental circuit can be used to demonstrate how optical communication systems operate.

The circuit is formed by a transmitter that uses a small laser pointer or a semiconductor laser module and a modulation system. Semiconductor lasers are powerful enough to send their signals over large distances (up to 1,000 ft) and can be obtained easily from many dealers. As a sensor, the receiver uses a photodiode or phototransistor that is sensitive enough to pick up the laser beam, even when it is located at a great distance from its source.

The project is experimental, as we use an empirical method to achieve modulation. But some suggestions are offered for readers who want to make further improvements in the circuit. With changes in the modulation system, using a Kerr cell for instance, the reader will be able to transmit high-frequency signals, including video or digital signals, from a PC serial port via the laser beam. It is also possible, by adding a multiplex circuit, to transmit several audio channels simultaneously using one laser beam.

How It Works

The light beam produced by a laser (red, for instance) can be used as a carrier to transport low-frequency signals such as audio, digital data, or video. Light is a form of electromagnetic waves, but it exhibits an ultra high frequency in the range of terahertz (trillions of hertz).

A laser is a special kind of light source. The radiation produced by a laser is monochromatic (i.e., it consists of only one frequency and is visible as a single color) and coherent. Because of these and other properties, a laser light beam is highly focused. The light does not spread out to form a large spot, even at great distances.

The concept here is to make a laser beam fall onto a small mirror that is placed on a loudspeaker's cone as shown in Fig. 5.29. When no signal is applied to the loudspeaker, the light beam can be picked up by a sensor placed as shown in the figure. But when an audio signal is applied to the loudspeaker, the cone's move-

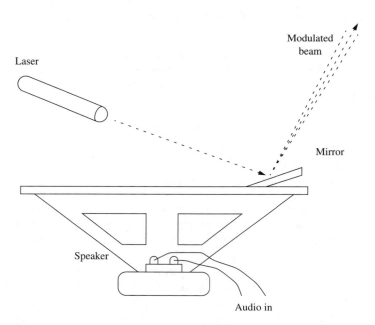

Figure 5.29   A loudspeaker used to modulate the laser beam.

ment is transferred to the mirror. As a result, the mirror oscillates according to the sound, changing the light beam reflection angle. The light beam oscillation is detected by the receiver, which generates a corresponding audio signal.

This adjustment is very critical because, if the light beam oscillates excessively, the sensor will pick up only the modulation peaks, chopping the signal. The experimenter must experiment to find the ideal volume applied to the loudspeaker that gives good reproduction in the receiver (see Fig. 5.30).

The experimental system can be built using any low-power amplifier (100 mW to 1 W) and a receiver. The receiver can be any common audio amplifier or the circuit suggested below.

This receiver circuit uses two complementary direct-coupled transistors and directly drives a small loudspeaker. The sensor is any phototransistor or photodiode. Using optical aids such as a convergent lens and a tube, the sensitivity can be increased and interference sources neutralized.

Assembly

The transmitter consists of any amplifier connected to a loudspeaker with a small mirror mounted on it. For the laser pointer, it is important to use an external power supply, as it will run a battery down in a short time. Figure 5.31 shows a 3 V power supply for the laser pointer and how the mirror can be affixed to a small loudspeaker to form the transmitter system. The ideal position for mirror placement must be determined experimentally. A diagram of the receiver is shown in Fig. 5.32. As this is an experimental circuit, it can be mounted on a solderless board or use a terminal strip as the chassis. Figure 5.33 shows the terminal strip version.

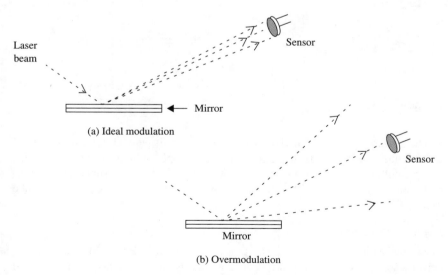

Figure 5.30   The applied audio level must be adjusted to avoid overmodulation.

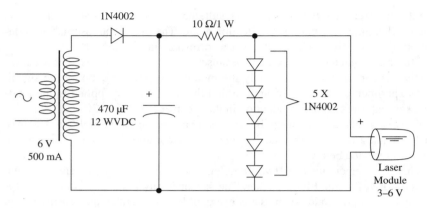

Figure 5.31 Power supply for a laser pointer module.

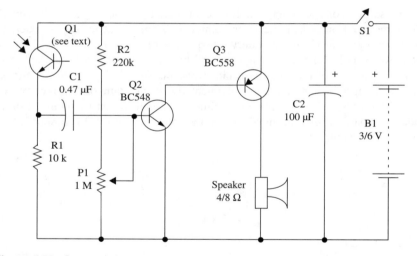

Figure 5.32 Laser receiver.

The power supply is formed by two or four AA cells, and P1 adjusts the gain of the circuit. The sensor is any phototransistor or photodiode. If you place this component inside a cardboard tube with a convergent lens, the sensitivity and directional characteristics will be increased. Any small 4 or 8 Ω loudspeaker can be used.

Adjustments and Use

Place the loudspeaker and the laser pointer as shown in Fig. 5.34. Place the receiver's sensor in a position to receive the laser beam. The sensor can be placed 10 ft or more away from the transmitter.

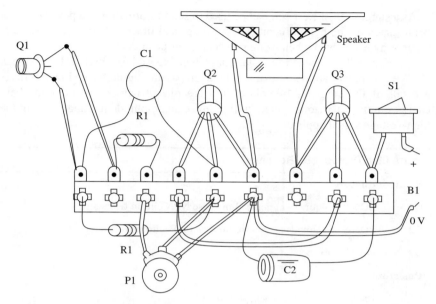

Figure 5.33   The circuit mounted on a terminal strip.

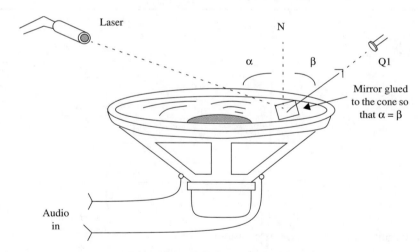

Figure 5.34   How to place the mirror and laser.

Turn on the amplifier and the receiver. As a signal source, you can use a CD player plugged into the transmitter input. Resistors Rx are important to avoid loading the output of the CD player. Observe that we are using the signal from the two channels at the same time, as they are mixed. Put the amplifier in a low-volume position and carefully adjust the transmitter's amplifier to get the best sound.

As a suggestion, a Kerr cell can be used to send information via powerful laser beams such as the helium-neon (HeNe) types and others. Ranges up to several miles can be achieved when using powerful laser beams.

The Kerr cell is a piece of crystal that, when exposed to a high-voltage signal, changes the polarization angle of a light beam passing through (see Fig. 5.35). As a result, if the elements of polarization are placed before and after the cell, changes in the polarization angle will produce an amplitude modulation with the high-voltage signal applied.

Parts List: Project 38 (Receiver)

Semiconductors

Q1   BC548 or equivalent general-purpose NPN silicon transistor

Q2   BC558 or equivalent genera-purpose PNP silicon transistor

Q3   any phototransistor

Resistors

R1   100,000 Ω—brown, black, yellow

R2   220,000 Ω—red, red, yellow

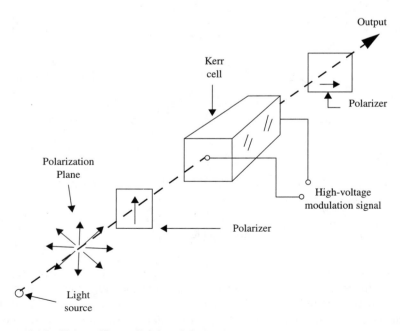

Figure 5.35 Using a Kerr cell for modulation.

Parts List: Project 38 (Receiver) (continued)

P1   1,000,000 Ω—potentiometer

Capacitors

C1   0.47 µF ceramic or metal film

C2   100 µF/6 WVDC electrolytic

Additional Parts and Materials

S1   SPST toggle or slide switch

B1   3 or 6 V, two or four AA cells

SPKR   small loudspeaker, 4 or 8 Ω

Terminal strip, battery holder, plastic box, cardboard tube and lens, wires, solder, etc.

Project 39: Video Transmitting Station

This transmitter can send the sound and video signals from a videocassette recorder or camera to a TV set placed up to 150 ft.

Features

- Range: 150 ft (typical)
- Power supply voltage: 117 Vac
- Frequency range: low VHF channels (2 to 6)
- Transmitted signal: composite video (audio, video, and synchronization)

Video cameras and videocassette recorders (VCRs) are equipped with audio and video outputs. The signals from these outputs can be used not only as a feed to a TV set, but as an input to a transmitter as well. This means that you can send the signals from a VCR in one room to a TV in another room without the use of cables.

The transmitter described here can send signals to distances up to 150 ft. Some suggested applications are as follows:

- It can be used for an experimental TV station in a school or at home, sending the programs to an adjacent room. A camera can be plugged into the circuit.
- You can distribute a videocassette program to TVs placed in several classrooms or different rooms in your home.
- It can be used for remote surveillance using a camera and the transmitter as shown in Fig. 5.36.
- For espionage, you can pick up both sounds and images using a remote camera and microphone.

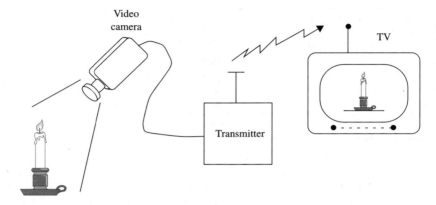

Figure 5.36   The circuit can be used for surveillance.

The signals can be received by a VCR as well as a TV, allowing you to record the scene. In the basic version, the circuit is powered from the ac power line but, for field applications, a 112 V battery can be used. It is also possible to change the high-frequency output stage to use high-power transistors such as the 2N2218. This will allow you to send signals to distances up to half a mile.

How It Works

The video signals need a wide transmission band as shown in Fig. 5.37. As shown, of the 6 MHz of bandwidth reserved for a TV channel, 4.5 MHz are occupied by video information and only a narrow band for the audio signal. A security band is reserved to separate the signals from adjacent channels.

As shown in the figure, the audio signal is displaced 4.5 MHz from the video signal carrier. This means that a transmitter designed to transmit both audio and

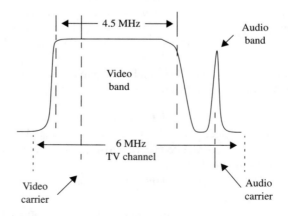

Figure 5.37   TV band.

video signals must be different from one that is designed to transmit audio only. The video channel is wider, which indicates the necessity of a modulation circuit that can operate with high-frequency signals. And at the same time, the circuit must operate with two modulation signals, one displaced 4.5 MHz from the other.

To perform those functions, our transmitter uses a configuration shown in the block diagram of Fig. 5.38. The main oscillator produces the high-frequency signal to be transmitted to the TV set. The circuit is formed by Q4, and the frequency is adjusted by CV1. This component should be adjusted to a free channel in the low VHF band between channels 2 and 6.

A BF494 is used for this circuit, but more powerful transistors such as the 2N2218 can be used. The signal produced by this stage passes through an amplification stage formed by Q3, and then through the final stage using transistor Q2. For more power, Q2 must be replaced by a 2N2218 and P1 replaced by a 100 Ω potentiometer. Note that this final amplification stage is *aperiodic*, which means that no adjustments should be made.

There are two modulation stages to be considered. The video signal is applied directly to the emitter of the last-stage transistor by P1. This trimmer potentiometer is used to find the best point of modulation according to the amplitude of the incoming signal. The audio signal must be displaced 4.5 MHz from this signal, and this is generated by another stage. A second oscillator using a BC548 transistor produces a 4.5 MHz signal modulated by the audio input.

The frequency for this stage is determined by L5 and the internal winding capacitance. It uses an adjustable core coil for this task, so the reader must be able

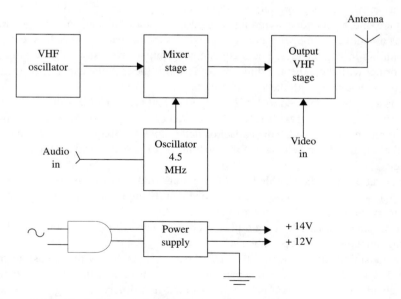

Figure 5.38 Block diagram of the video transmitting station.

to adjust this circuit to displace the audio signal exactly 4.5 MHz from the video signal. When the 4.5 MHz audio modulated signal beats with the center of the video carrier given by Q4, it is displaced by 4.5 MHz, producing the desired high-frequency TV signal. The audio signal is applied to the second high-frequency stage. P2 is used to control the audio modulation.

Notice that the chroma, synchronization, and brightness signals are present in the composite video signal, so the transmitter can operate with PAL, NTSC, or any other common video systems. You only need to ensure that the receiver (TV) can recognize the video signal.

The power supply has two outputs. One of them produces 14 Vdc voltage and other a 12 Vdc voltage. The 12 Vdc output is regulated by a zener diode.

Observe that the circuit has an external power supply input. Any 12 Vdc source can be plugged into this point to power the circuit, including batteries, cells, and other sources.

Assembly

The schematic diagram of the video transmitting station is shown in Fig. 5.39. The circuit is mounted on a printed circuit board as shown in Fig. 5.40.

The electrolytic capacitors must be rated to 16 WVDC or more. Capacitors with values below 1 µF can be ceramic or another type according to parts list specifications.

L1 and L3 are formed by four turns of AWG 22 enameled wire on a 3 mm dia. form with no core. L2 is formed by three turns of AWG 22 enameled wire on a 3 mm dia. form with no core.

Transformer T1 has a primary winding formed by 40 turns of AWG 32 enameled wire and a secondary winding formed by 10 turns of the same wire. The form is a 4 mm dia. tube with a ferrite core. L5 is the primary and L6 the secondary winding in the diagram. The form can be obtained from any TV or radio FI transformer with the indicated dimensions. Remove the metallic cap and the original coil, replacing it with the specified coils.

D1 is a zener diode rated to 4.7 V × 400 mW, and D2 is a 12 V × 400 mW zener. You can experiment with values of C8. Values in the range between 22 and 100 pF are recommended. This capacitor determines the frequency of the audio modulation, and it can be changed if adjusting of T1 doesn't produce the correct tuning.

Input cables must be shielded, and RCA plugs and jacks are used to connect the signal sources. It is important to use good quality video cables, and keep them short to avoid unsatisfactory images in the receiver.

The antenna is a length of rigid plastic-covered wire 12 to 20 in. long. A telescoping antenna is also acceptable.

The transmitter should be installed in a plastic or metal box with dimensions of about 5 × 4 × 2 in. Screws and separators are used to mount the printed circuit board inside the box. T1 is a power transformer. Its primary winding is rated to the local ac power line voltage and the secondary winding to 9 V × 150 mA.

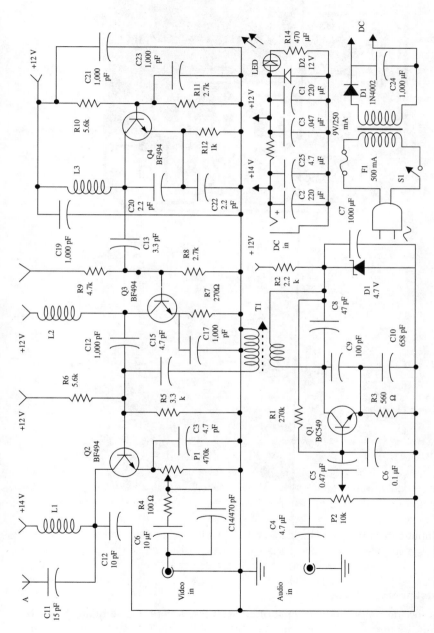

Figure 5.39   Video transmission station.

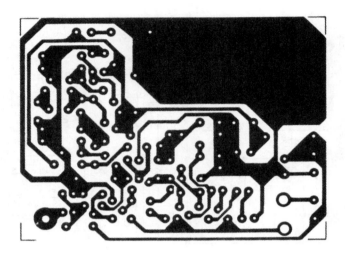

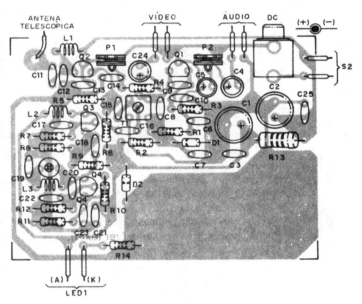

Figure 5.40   Printed circuit board for video transmission station.

Adjustments and Use

Plug the transmitter to the output of a videocassette with a tape running. Notice that the inputs for audio and video are different. Place a TV close to the transmitter and tune the TV to channel 3 or 4 (whichever is free in your location). A portable TV with an internal antenna is recommended to make these adjustments.

Power the transmitter up and first adjust CV to pick up the transmitted signal. Next, adjust P2 to get the best image in the TV set. Be careful not to tune a harmonic. Ensure that you have a good contrast. If necessary, retune CV.

Now, the reader must move to the audio modulation adjustments. Adjust T1, moving the core with a plastic or wooden tool until the sound is reproduced in the TV. If this does not occur, you must alter the value of C8. At the same time, adjust P2 to obtain the distortion-free sound reproduction.

When using the transmitter, keep the antenna in the vertical position and far away from metallic objects that can affect signal propagation. Large walls that contain metallic structures can negatively affect signal propagation, reducing range. If distortions appear (e.g., a wavy image), you may need to add power supply filtering, or there may be problems with the cables.

To use the circuit with a video camera, plug the video and audio outputs into the transmitter. When using this transmitter, remember that your neighbors can also pick up the transmitted images in their TV sets.

Parts List: Project 39

Semiconductors

Q1	BC548 or equivalent general-purpose NPN silicon transistor
Q2, Q3, Q4	BF494 or equivalent high-frequency NPN silicon transistor
D1	4.7 V × 400 mW zener diode
D2	12 V × 400 mW zener diode, 1N4742
D3	1N4002 silicon rectifier diode
LED1	common red LED

Resistors (1/8W, 5%)

P1	470 Ω—trimmer potentiometer
P2	10,000 Ω—trimmer potentiometer
R1	270,000 Ω—red, violet, yellow
R2	2,200 Ω—red, red, red
R3	560 Ω—green, blue, brown
R4	100 Ω—brown, black, brown
R5	3,300 Ω—orange, orange, red
R6, R10	5,600 Ω—green, blue, red
R7	270 Ω—red, violet, brown
R8, R11	2,700 Ω—red, violet, red
R9	4,700 Ω—yellow, violet, red
R12	1,000 Ω—brown. black, red

Parts List: Project 39 (continued)

R13	47 Ω/1 W—yellow, violet, black
R14	470 Ω—yellow, violet, brown

Capacitors

C1, C2	220 µF/16 WVDC electrolytic
C3, C25	4,700 pF ceramic
C4	4,7 µF/16 WVDC electrolytic
C5	0.47 µF ceramic or electrolytic
C6, C9	100 pF ceramic
C7, C17, C19, C21, C23	1,000 pF ceramic
C8	47 pF ceramic
C10	68 pF ceramic
C11	15 pF ceramic
C12	10 pF ceramic
C13, C15	4.7 pF or 5 pF ceramic
C14	470 pF ceramic
C16	15 pF ceramic
C18	2.7 or 3 pF ceramic
C20, C22	2.2 or 2 pF ceramic
C24	1,000 µF/16 WVDC electrolytic
C26	10 µF/16 WVDC electrolytic

Additional Parts and Materials

T1	transformer (see text)
T2	transformer: primary winding rated to the local power line voltage and secondary winding rated to 9 V × 250 mA
S1	SPDT toggle or slide switch
S1	SPST toggle or slide switch
L1, L2, L3, L4	coils (see text)
A	antenna (see text)
J1, J2	RCA jacks and plugs (see text)

Printed circuit board, plastic or metallic box, power cord, shielded cable, wires, etc.

6

Miscellaneous Projects

The following projects are intended to help the reader in his efforts to build transmitters and associated circuits, improve their performance, and power them from the ac line. It also offers solutions for making adjustments and testing parts.

Project 40: Field Strength Meter

This device can indicate if a transmitter or high-frequency oscillator stage is operating. Powered from AA cells or a battery, the circuit can detect signals in the range between 500 kHz and 200 MHz.

Features

- Power supply voltage: 3 to 9 V
- Current drain: 200 μA (typical)
- Frequency range: 500 kHz to 200 MHz

How can you know if a transmitter is generating a radio signal when it cannot be located by tuning the receiver? Is the transmitter dead, or are the signals being produced within an unexpected frequency range?

Using a field strength meter, the reader can detect the signals produced by any oscillator running between 500 kHz and 200 MHz. If the signal is detected but not tuned by the receiver, this will reveal that the problem is a frequency adjustment and not an operational problem. By changing the coils or making further adjustments, it is possible to correct the problem.

The field strength meter described here is very simple and uses only one transistor, but it is sensitive enough to pick up the weak signals produced by small transmitters as described in this book. In addition, it can be used for such other tasks as

- Getting better performance when adjusting or locating antennas
- Finding hidden transmitters that have been concealed by spies
- Finding places where the signals from a transmitter are blocked by obstacles

How It Works

A piece of rigid wire or a telescoping antenna is used to pick up the radio signals. Very low-frequency signals, such as hum from the ac power line and others, are shunted to ground by L1. High-frequency signals (500 kHz and above) are applied to D1. These high-frequency signals are detected and then applied by C1 to the base of a transistor. The transistor acts as an amplifier, driving a meter.

The higher the intensity of the detected signal, the higher the current through the meter. Therefore, when the circuit picks up a strong signal, a proportionally strong current flows through the transistor and is displayed by the indicator (meter).

As this circuit is not tuned to any particular frequency, all ambient signals can be picked up simultaneously. The circuit is very sensitive thanks to the use of a germanium diode, and the only required adjustment is made using P1 to set the transistor close to the point at which it begins to conduct.

Assembly

Figure 6.1 shows the complete diagram of the field strength meter. As the circuit is not critical, the components can be mounted on a terminal strip as shown in Fig. 6.2. The reader who prefers a compact unit can etch a small printed circuit board for mounting the components.

XRF is any 47 to 470 µH micro-coil. The reader can use a homemade version of this component. It is sufficient to wind 100 to 200 turns of AWG 32 enameled wire on a 100,000 Ω × 1/4 W resistor. For the readout, any microammeter with scales between 100 and 500 µA can be used. The circuit can be powered from a 3 to 9 V power supply and installed in a plastic box. The dimensions of the plastic box are basically determined by the size of the power supply.

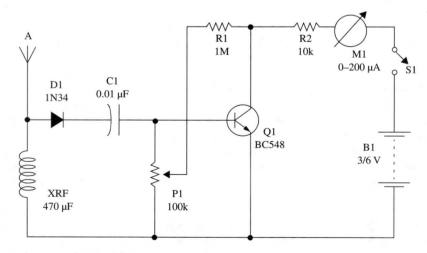

Figure 6.1   Field strength meter.

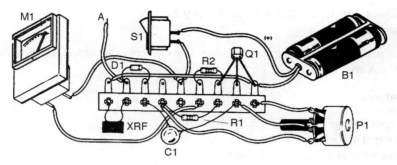

Figure 6.2   The project can use a terminal strip as chassis.

Testing and Use

Use S1 to power the circuit up, and adjust P1 to set the needle to the bottom of the meter's scale (near zero). Then, using any transmitter as a signal source, place the field strength meter near it. As you move the field strength meter closer to the transmitter's antenna, the meter will indicate the presence of a signal. The closer the meter is to the transmitter, the higher the indicated signal strength.

Never touch the field strength meter's antenna directly to the transmitter's antenna or output stage. The strong signal can damage the device.

Parts List: Project 40

Semiconductors

Q1   BC548 or equivalent general-purpose NPN silicon transistor

D1   1N34, 1N60, or equivalent—any germanium diode

Resistors (1/8 W, 5%)

R1   1,000,000 Ω—brown, black, green

R2   10,000 Ω—brown, black, orange

P1   100,000 Ω potentiometer

Capacitors

C1   0.01 µF ceramic

Additional Parts and Materials

XRF   47 µH to 470 µH RF choke (see text)

M1   0–200 µA microammeter

S1   SPST toggle or slide switch

B1   3 to 9 V, AA cells or battery

Terminal strip or printed circuit board, plastic box, battery holder or clip, antenna, knob for the potentiometer, wires, solder, etc.

Project 41: 12 V × 5 A Power Supply for Transmitters

This power supply can be used to power transmitters with current drains between 500 mA and 5 A.

Features

- Input voltage: 117 Vac (or 220/240 Vac)
- Output voltage: 12 V
- Output current: 5 A (maximum)

A 12 Vdc × 5 A power supply such as described here can be very useful to the reader who builds transmitters. But it is also beneficial if you want to use the ac power line as a supply for devices that normally are powered by cells or your car's battery. These include CD players, tape recorders, AM/FM radios, and so on. Of course, with a current rating of about 5 A, this power supply can be used with even the most powerful transmitters described in this book. In any event, it provides a good way to save money on expensive cells and batteries.

How It Works

The transformer reduces the ac power line voltage to 12 V. This voltage is rectified by a full-wave configuration using two diodes. Filtering is provided by C1.

Regulation is the job of Q1 and IC1. IC1 is a three-terminal voltage regulator IC with a maximum output of 1 A. To boost this current, a PNP transistor is used, increasing the maximum output to 5 A. This PNP transistor is wired so that its output voltage is determined by the IC. Both the transistor and IC must be mounted on heat sinks.

As high current flows through the circuit, the reader must be careful to use appropriate wiring. Make sure that the wires are capable of carrying this circuit's maximum output.

The circuit also includes a fuse to protect it. You can also add an LED to indicate when the power is on. This LED can be wired in series with a 2,200 Ω × 1/2 W resistor, with both in parallel with C1.

Assembly

A complete diagram of the power supply is shown in Fig. 6.3. As the circuit is not critical, a terminal strip can be used as a chassis for mounting the principal components. Note that the transformer must be fixed in the box, as are the fuse holder and the power switch. The power supply is mounted as shown in Fig. 6.4.

The output can be accessed via terminals, which is common for this kind of device. Red and black terminals are recommended for easy identification of the positive and negative poles.

The transistor and the integrated circuit must be mounted on heat sinks. The heat sink for the transistor must be larger than the one used for the transistor, as is

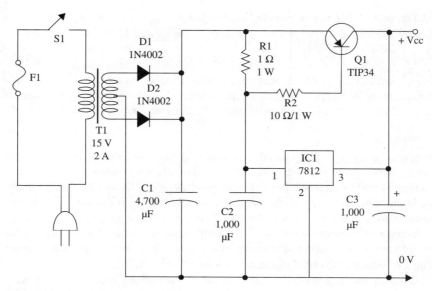

Figure 6.3 Power supply for transmitters.

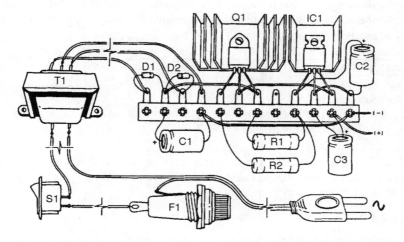

Figure 6.4 Power supply mounted using a terminal strip as the chassis.

evident in the figure. This is necessary because the transistor conducts 5 A, but the IC carries only 1 A, when the load's current flow is at maximum.

Any diode rated to 50 V (PIV) × 5 A can be used for this project. The capacitor (C1) can have a capacitance rating between 4,700 and 10,000 μF, and it should be rated to voltages of 25 WVDC or more. The other electrolytic capacitors need to be rated to 16 WVDC or more.

The transformer has a primary winding that is appropriate for your local ac power line voltage. The secondary winding can be rated for voltages between 12 V CT and 15 V CT and currents of 5 A.

All of the components can be installed in a plastic or metal box. If a metal box is your preference, it can be used as a ground connection for the dc voltage. But be careful with components that are powered directly from the ac power line, as any isolation fault can cause severe shock hazards or dangerous shorts.

Testing and Use

Power on the circuit after installing a fuse in the fuse holder. To the output, connect a 12 V × 200 mA to 1 A incandescent lamp (such as used in cars). Using a multimeter adjusted to the 0–15 dc voltage scale, see if the output voltage is correct. The value must be in the range between 12 and 12.8 V, and the lamp must glow as expected.

When using the device, observe the proper polarity when connecting any device to this power supply. Don't try to use it to power any device that requires currents beyond this supply's maximum output.

If the transistor tends to overheat and the output voltage falls below 10 V when supplying a particular device, turn off the power supply immediately. This is an indication that the device drains more current than the power supply can provide, or that there is a problem related to a short circuit. If hum or instability problems occur when using this device to power a transmitter, wire a 0.1 µF ceramic capacitor in parallel with the output.

Parts List: Project 41

Semiconductors

IC1   7812 voltage regulator, integrated circuit

Q1   TIP34 high-power PNP silicon transistor

D1, D2   any 50 V/5 A silicon rectifier diodes

Resistors

R1   1 Ω × 2 W wirewound

R2   10 Ω × 2 W × wirewound

Capacitors

C1   4,700 µF × 25 WVDC electrolytic

C2   1,000 µF × 25 WVDC electrolytic

C3   100 to 1,000 µF × 16 WVDC electrolytic

Parts List: Project 41 (continued)

Additional Parts and Materials

S1   SPST toggle or slide switch

T1   transformer: primary winding according to the ac power line; secondary
            winding 15 V CT × 1 A (see text)

F1   1 A fuse and holder

Terminal strip, plastic or metallic box, power cord, output terminals, heat sinks for
IC1 and Q1, wires, solder, etc.

Project 42: Dip Meter

This circuit is ideal for the analysis of tuned or resonant LC circuits. Anyone who
builds transmitters needs a device like this.

Features

- Frequency range: 10 to 200 MHz
- Power supply voltage: 9 Vdc
- Number of transistors: 1

This instrument is extremely useful for any reader who works with high-frequency circuits or tuned circuits such as those found in transmitters. Using this
instrument, it is possible to determine the resonant frequency of tuned circuits.

The classic version of this instrument was called a *grid dip meter*, as it used a
tube in a configuration such as shown in Fig. 6.5. A state of resonance would
cause the grid current of the tube to drop, creating a "dip" that can be easily seen
by means of a meter wired to this element.

The circuit was formed by a tube in an oscillator configuration with a meter
connected to its grid. When its coil was placed near a resonant circuit, an energy
transfer between the circuits affected the grid current, which was indicated on the
meter. Observe that this energy transfer, as indicated by the meter, would occur
only if the circuits were tuned to the same frequency. So, using a device like this,
it was sufficient to place it near the resonant circuit and check the meter. If the
meter indicated a change in the current, then this was an indication that the oscillator frequency and the resonant frequency of the circuit under analysis were the
same.

To be useful, the device was based on variable oscillator with a calibrated scale
from which the frequency could be read. Therefore, to determine the frequency
of any resonant circuit, the device was placed near the circuit, and the frequency
was adjusted to the point at which the meter indicated a current change. The frequency then would be shown on the meter.

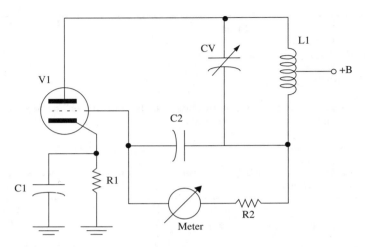

Figure 6.5   A "grid dip meter" using tube.

Today, the same type of device can be made using modern components such as bipolar transistors, field effect transistors, and others. Of course, these components don't have a grid, so the word has been dropped from the instrument's name.

A dip meter using the BF494 transistor is shown in Fig. 6.6. This circuit can run at frequencies up to 200 MHz. Equivalent transistors such as the 2N2222 and others with higher cutoff frequencies can be used. To operate in the frequency range between 70 and 110 MHz, the coil is formed by four turns of 18 to 22 enameled wire wound on a coreless 1 cm dia. form. Figure 6.7 shows how the circuit can be assembled using a plastic box.

Note that the coil is plugged into a socket that is installed in the external wall of the box. The reader can wind some coils with different numbers of turns for operation in different frequency ranges. Of course, the variable capacitor must be calibrated to each of the coils, with different scales. A reference oscillator (signal generator) can be used for this task.

To use the dip meter, you just need to move the meter's coil close to the tuned circuit under analysis. Use the trimmer potentiometer to adjust the meter so the needle does not move beyond the end of the scale. While varying the CV adjustment and observing the meter needle, it is possible to locate a point at which the current flow changes. The frequency will be shown on the meter.

Figure 6.8 shows another version of a solid state dip meter using an FET. This circuit runs in a range of frequencies between 100 kHz and 100 MHz, depending on the coil used. For 100 kHz signals, the coil is formed by 200 + 200 turns of AWG 28 enameled wire on a 1 in. (2.5 cm) dia. PVC tube. For the 500 kHz range, the coil is formed by 60 + 60 turns of the same wire wound on the same tube. For the 2 MHz range, the coil is formed by 30 + 30 turns, and for 5 MHz by

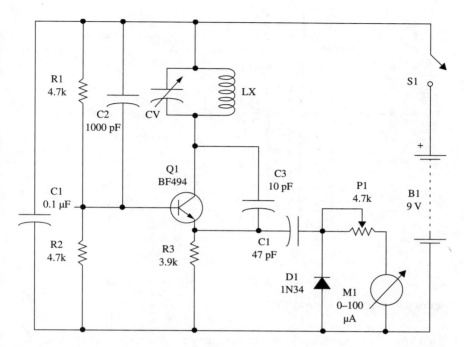

Figure 6.6   A "grid dip meter" using tube.

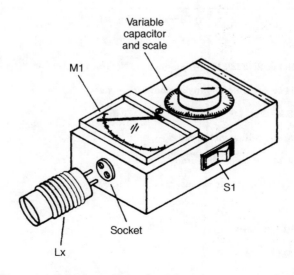

Figure 6.7   Dip meter mounting using a plastic box.

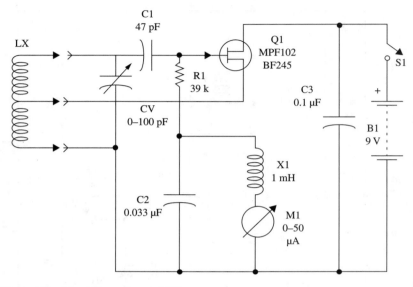

Figure 6.8   Dip meter using an FET.

15 + 15 turns. For 40 MHz, the coil is formed by 5 + 5 turns. For all of these, the tube is the same, and no ferrite core is necessary. If properly adjusted, the dip meter can replace a frequency meter in many applications.

Parts List: Project 42

(a) Circuit of Fig. 6.6

Semiconductors

Q1   BF494 or 2N2222 high-frequency NPN silicon transistor

D1   1N34 or equivalent—any germanium diode

Resistors (1/8 W, 5%)

R1, R2   4,700 Ω—yellow, violet, red

R3   3,900 Ω—orange, white, red

P1   47,000 Ω—trimmer potentiometer

Capacitors

C1   0.1 μF ceramic

C2   1,000 pF ceramic

C3   10 pF ceramic

Parts List: Project 42 (continued)

C4   47 pF ceramic

Additional Parts and Materials

LX   coil (see text)

CV   Variable, 120 to 290 pF

M1   0–50 µA microammeter

S1   SPST toggle or slide switch

B1   9 V battery

Printed circuit board, battery clip, plastic box, forms for coils, knob to the variable capacitor, wires, solder, etc.

(b) Circuit of Fig. 6.8

Semiconductors

Q1   BF245 or MPF102 junction field effect transistor (JFET)

Resistor (1/8W, 5%)

R1   39,000 Ω—orange, white, orange

Capacitors

C1   47 pF ceramic

C2   0.033 µF ceramic

C3   0.1 µF ceramic

CV   100 to 190 pF variable capacitor

Additional Parts and Materials

M1   0–50 µA microammeter

X1   1 mH RF choke

S1   SPST toggle or slide switch

LX   coils (see text)

B1   9 V battery

Printed circuit board, plastic box, knob with scale for the variable capacitor, socket for the coil, battery clip, wires, etc.

Appendix
Useful Information for
Transmitter Construction and Use

A.1 Morse Code

A ·−	S ···
B −···	T −
C −·−·	U ··−
D −··	V ···−
E ·	W ·−−
F ··−·	X −··−
G −−·	Y −·−−
H ····	Z −−··
I ··	1 ·−−−−
J ·−−−	2 ··−−−
K −·−	3 ···−−
L ·−··	4 ····−
M −−	5 ·····
N −·	6 −····
O −−−	7 −−···
P ·−−·	8 −−−··
Q −−·−	9 −−−−·
R ·−·	0 −−−−−
Comma −−··−−	Question mark ··−−··
Error ·······	Wait ·−···
Period ·−·−·−	Double dash −···−
Fraction bar −··−·	Invitation to transmit −·−
End of message ·−·−·	End of transmission ···−·−

A.2   Resistor Color Code

Black	0
Brown	1
Red	2
Orange	3
Yellow	4
Green	5
Blue	6
Violet	7
Gray	8
White	9

A.3   TV and Other Broadcast Frequencies

MW Broadcasting

Frequency range	535 to 1,605 kHz
Number of channels	107
Channel width	10 kHz

FM Broadcasting

Frequency range	88 to 108 MHz
Number of channels	100
Channel width	200 kHz

VHF TV, Low Channels

Frequency range	54 to 88 MHz
Number of channels	5 (2 through 6)
Channel width	6 MHz

VHF TV, High Channels

Frequency range	174 to 216 MHz
Number of channels	7 (7 through 13)
Channel width	6 MHz

UHF TV

Frequency range	470 to 890 MHz
Number of channels	70 (14 through 83)
Channel width	6 MHz

A.4 Wavelengths/Bands

Band	Frequency range	Metric Name	Designation
4	3 to 30 kHz	Myriametric waves	VLF
5	30 to 300 kHz	Kilometric waves	LF
6	300 to 3,000 kHz	Hectometric waves	MF
7	3 to 30 MHz	Decametric waves	HF
8	30 to 300 MHz	Metric waves	VHF
9	300 to 3,000 MHz	Decimetric waves	UHF
10	3 to 30 GHz	Centimetric waves	SHF
11	30 to 300 GHz	Millimetric waves	EHF
12	300 to 3,000 GHz	Decimillimetric waves	–

A.5 Bare Annealed Copper Wire

The table below lists American Wire Gauge (AWG) numbers ranging from 10 to 40. The larger the gauge number, the greater its current-carrying capacity.

Gauge no.	Diameter (mils)	Gauge no.	Diameter (mils)
10	101.90	26	15.94
11	90.74	27	14.20
12	81.81	28	12.64
13	71.96	29	11.26
14	64.08	30	10.03
15	57.07	31	8.93
16	50.82	32	7.95
17	45.26	33	7.08
18	49.30	34	6.31
19	35.89	35	5.62
20	31.96	36	5.00
21	28.46	37	4.45
22	25.35	38	3.97
23	22.57	39	3.53
24	20.10	40	3.15
25	17.90		

A.6   Propagation Characteristics of Radio Waves

VLF: 20–30 kHz.   VLF is very stable with low attenuation at all times. It is influenced by magnetic storms produced by the sun. Ground waves extend over long distances, allowing continuous communications.

LF: 30–300 kHz.   LF displays some seasonal and daily variations. Daytime absorption is greater than VLF, increasing with frequency. It is used for long distance communications.

MF: 300–3,000 kHz.   MF attenuation is low at night and high in the daytime; it has greater gain in summer than in winter. The skywaves can be reflected. Ground attenuation is high over land and low over salt water.

HF: 3–30 MHz.   HF depends on ionospheric conditions. Considerable variations exist from day to night and from season to season. It is used for medium and long distance communications.

VHF: 30–300 MHz.   VHF sometimes depends on ionospheric conditions. Variations occur from season to season and, during the day, the signals are reflected by the ionosphere. Quasi-optical transmission performance is evident. It is used for medium- and long-distance communications.

UHF: 300–3,000 MHz.   UHF is the same as the VHF. Under abnormal conditions, it can be reflected by the ionosphere. It is used for short-range communications.

SHF: 3–30 GHz.   SHF is the same as UHF. Water vapor can absorb the signals. It is used for short-range communications.

A.7   Names of the Frequency Bands

VLF	20–30 kHz	Very low frequency
LF	30–300 kHz	Low frequency
MF	300–3,000 kHz	Medium frequency
HF	3–30 MHz	High frequency
VHF	30–300 MHz	Very high frequency
UHF	300–3,000 MHz	Ultra high frequency
SHF	3–30 GHz	Super high frequency

A.8   International Phonetic Alphabet

Letter	Name	Pronunciation
A	Alpha	AL-fah
B	Bravo	BRAH-voh
C	Charlie	CHAR-lee
D	Delta	DELL-tah
E	Echo	ECK-oh
F	Foxtrot	FOKS-trot
G	Golf	GOLF
H	Hotel	HOH-tel
I	India	IN-dee-ah
J	Juliet	JEW-lee-ett
K	Kilo	KEY-loh
L	Lima	LEE-mah
M	Mike	MIKE
N	November	No-VEM-ber
O	Oscar	OSS-cah
P	Papa	Pah-PAH
Q	Quebec	Keh-BECK
R	Romeo	ROW-me-oh
S	Sierra	See-AIR-rah
T	Tango	TANG-go
U	Uniform	YOU-nee-form or OO-nee-form
V	Victor	VIK-tah
W	Whiskey	WISS-key
X	X-ray	ECKS-ray
Y	Yankee	YANG-key
Z	Zulu	ZOO-loo

A.9   Component Dealers

Finding parts for the described projects is not easy if the reader doesn't live near a very good component shop. For the readers who live in distant places, the postal service provides an important alternative for parts acquisition.

Some important dealers around the world sell by mail and also have Internet sites where their online catalogs are accessible. Some of these dealers are indicated below.

U.S.A.

ALL ELECTRONICS
P.O. Box 567
Van Nuys, CA 91408-0567
Internet: http://www.allcorp.com

C&S SALES
150 W Carpenter Avenue
Wheeling, IL 60090
Internet: http://www.elenco.com/cs_sales

MOUSER
958 North Main St.
Mansfield, TX 76063
Internet: http://www.mouser.com

RADIO SHACK
500 One Tandy Center
Forth Worth, TX 76102
Internet: http://www.radioshack.com

FOREIGN

RS COMPONENTS
P.O. Box 99
Corby Northants
NN17 9RS United Kingdom
Internet: http://www.worldserver.pipex.com/rs/index
Offices: Australia, Denmark, Hong Kong, Italy, Singapore, Sweden, Austria, France, India, Malaysia, South Africa, United Kingdom, Brazil, Chile, Germany, Republic of Ireland, New Zealand, and Spain.

MAPLIN ELECTRONICS
P.O. Box 3
Rayleigh
Essex SS6 2BR United Kingdom
Internet: http://www.qkit.com.public/qkits/products.htm